Atlas of Neolithic plant remains from northern central Europe

Advances in Archaeobotany

Volume 4

Website
www.archaeobotany.eu

ISSN
2405-7452

Contact the Publisher
Barkhuis
Kooiweg 38
9761 GL Eelde
The Netherlands
info@barkhuis.nl
www.barkhuis.nl

Atlas of Neolithic plant remains from northern central Europe

Wiebke Kirleis

Barkhuis Publishing
Groningen 2019

Book design: Agnes Heitmann, Janine Cordts, UFG Kiel
Cover design: Agnes Heitmann, UFG Kiel
Copy editor: Suzanne Needs-Howarth, Toronto

Cover image: Charred plant assemblage from the well feature, Oldenburg LA77, Dannau (section 5.1, photo: Wiebke Kirleis, UFG Kiel)

The publication was subsidized by German Research Foundation (Deutsche Forschungsgemeinschaft, DFG) through the collaborative research programs SPP1400 'Early Monumentality and Social Differentiation', SFB1266 'Scales of Transformation: Human–Environmental Interaction in Prehistoric and Archaic Societies' (Project Ref. 2901391021), and the Botanical Platform of the Graduate School 'Human Development in Landscapes' (GSHDL GSC 208).

ISBN 9789492444912

This atlas is dedicated with gratitude to Dušanka Kučan, of Wilhelmshaven, Germany, for introducing me to the wonderful and aesthetic world of macro-photography and for teaching me to get intensively involved with the plant remains, to esteem every single seed or fruit, whether tiny or huge, and to stage and illuminate each plant remain as if it were an 'archaeobotanical shooting star'.

Content

Preface: The atlas of Neolithic plant remains from northern central Europe

This monograph presents examples of ancient plant remains from 36 Neolithic sites that were investigated mainly in the framework of three collaborative research programs at Kiel University, SPP1400 'Early Monumentality and Social Differentiation', SFB1266 'Scales of Transformation: Human–Environmental Interaction in Prehistoric and Archaic Societies', and the botanical platform of the Graduate School 'Human Development in Landscapes' (GSHDL), which complements ongoing PhD projects with archaeobotanical studies. The archaeology of each site is briefly introduced in order to provide the context for the ancient plant assemblages, from which selected remains are presented on accompanying plates, in the form of macro-photographs. The material presented here allows a first glimpse into the composition and the preservation of the plant assemblages from these 36 sites. Archaeobotanical find lists and radiocarbon dates for each site, as well as an index of the documented species, are provided in the appendices.

The digital photographs were taken with a Leica Macrophot Z6 Apo combined with an Olympus Digital Camera UC30 and the Software package CellSens, at the Institute for Prehistoric and Protohistoric Archaeology at Kiel University. Mass finds and large remains were documented with the use of a Nikon D750. To gain a high depth of focus, in most cases, multiple photographs were taken per set and then combined, using the stacking program Helicon Focus 6. The photographs for the plates in section 2.1, for the site of Rosdorf-Mühlengrund, were taken at the Niedersächsisches Institut für historische Küstenforschung (Lower Saxony Institute for Historical Coastal Research), under the guidance of Dušanka Kučan, with an analogue Leica M1 camera mounted on a Leitz Aristophot.

Multiple people have contributed to the success of this volume. I am grateful to numerous institutions and persons for permission to print photographs and drawings from excavations and excavation reports and for various other types of support.

Archäologisches Museum Hamburg: Jochen Brandt; Archäologisches Landesamt Schleswig-Holstein (ALSH), Schleswig: Birte Anspach, Anja Austen, Annette Guldin, Steffen Haucke, Veronika Klems, Stefanie Klooß, Ingo Lütjens, Eicke Siegloff; Bournemouth University: Timothy Darvill; Brandenburgisches Landesamt für Denkmalpflege und Archäologisches Landesmuseum (BLDAM), Wünstorf: Susanne Jahns; Christian-Albrechts-Universität Kiel (CAU), Kiel: Susanne Beyer, Jan-Piet Brozio, Janine Cordts, Marta Dal Corso, Yasmin Dannath, Walter Dörfler, Ingo Feeser, Dragana Filipović, Mark Hadyniak, Agnes Heitmann, Sara Jagiolla, Helmut Kroll, Ben Krause-Kyora, Ricarda Liedtke, John Meadows, Johannes Müller, Ralf Opitz, Tanja Reiser, Anna E. Reuter, Edeltraud Tafel, Richard Bleckmann, Karin Winter; Deutsches Archäologisches Institut (DAI), Berlin: Anja Behrens, Reinder Neef; Niedersächsisches Landesamt für Denkmalpflege, Stützpunkt Oldenburg: Jana Esther Fries, Jürgen Schneider; Moesgaard Museum, Højbjerg, Denmark: Niels H. Andersen; Niedersächsisches Institut für historische Küstenforschung (NIhK), Wilhelmshaven: Moritz Mennenga; WHP Archäologiebüro GbR: Andrea Hahn-Weishaupt; Zentrum für Baltische und Skandinavische Archäologie (ZBSA), Schleswig: Berit Eriksen.

The German Research Foundation (Deutsche Forschungsgemeinschaft, DFG) is deserving of the utmost thanks for providing the funding for SPP1400 'Early Monumentality and Social Differentiation'; SFB1266 'Scales of Transformation: Human–Environmental Interaction in Prehistoric and Archaic Societies' (Project Ref. 2901391021); and the Graduate School 'Human Development in Landscapes' (GSHDL GSC 208). I personally am much obliged to the Niedersächsisches Ministerium für Wissenschaften und Kultur for funding the project on the archaeobotanical investigations on the LBK site of Rosdorf-Mühlengrund at Göttingen University.

1 Introduction

Our knowledge about prehistoric plant use is based on the materiality of ancient plant remains as uncovered by archaeological excavations and on their study within the academic field of palaeo-ethnobotany. Research is centred around the archaeobotanical identification of plant remains that have survived for thousands of years in the ground or that have left imprints in ceramics and daub. Crucial for the proper identification of ancient, fossilized plant remains is the preservation of the morphological and anatomical characteristics of seeds, fruits, tubers, and other plant parts. It allows for comparison of ancient plant remains, such as those documented here, with modern remains, such as those presented e.g. in 'Digital seed atlas of the Netherlands' (Cappers et al., 2006, now in its second edition [2012]), 'Digital atlas of economic plants' (Cappers et al., 2009), and 'Digital atlas of economic plants in archaeology' (Neef et al., 2012). In this volume, mainly charred and in some instances

Fig. 1.1. Map showing the geographical distribution of the 36 archaeological sites with ancient plant assemblages presented in this volume, with the respective section numbers.

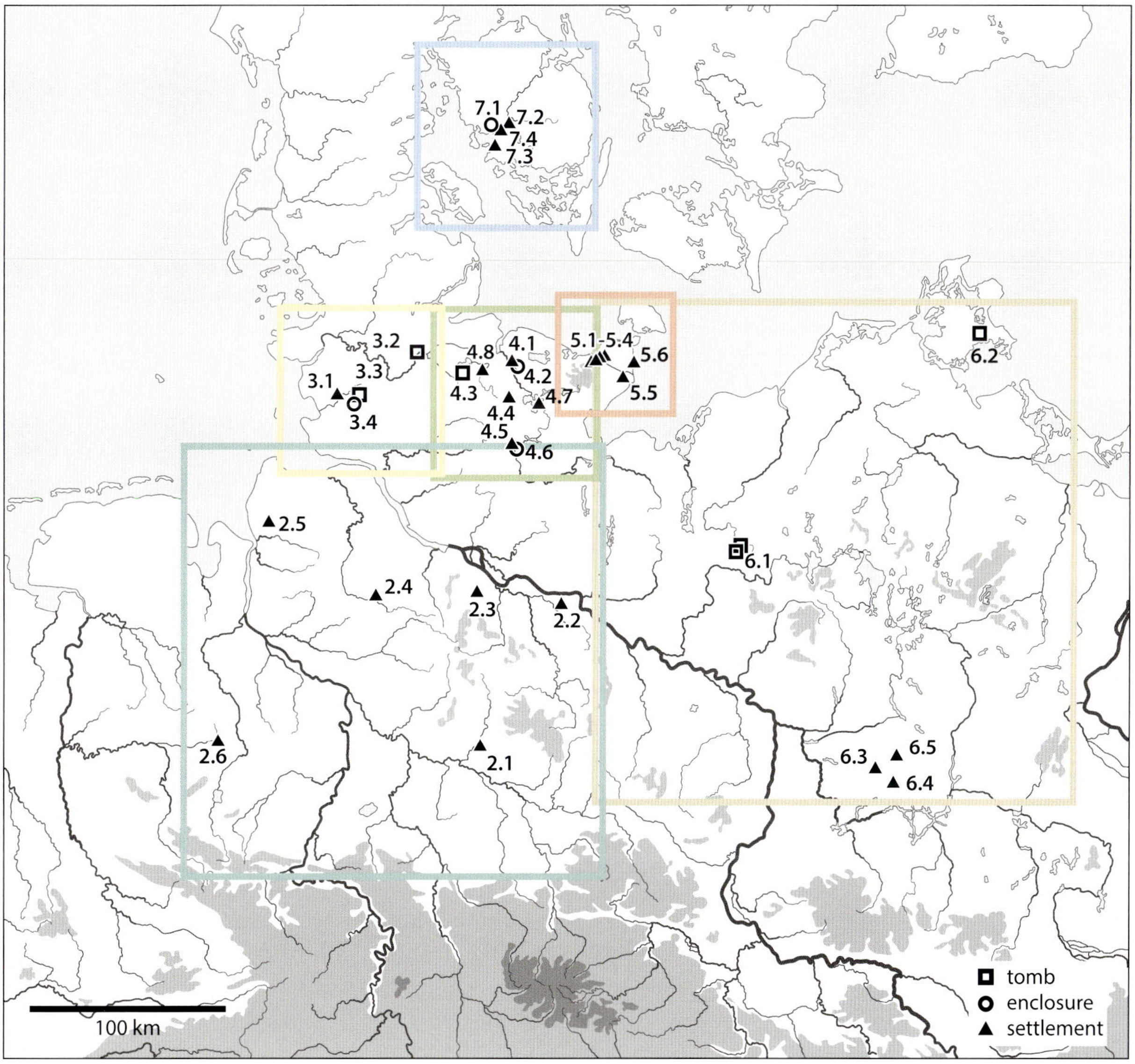

subfossil plant remains from 36 sites in central and northern Germany and the island of Funen, Denmark, are presented (Fig. 1.1 and Fig. 1.2) to show the comparability as well as the variation in the traits of these remains for the Neolithic period (chronology in Fig. 1.3). The traditional nomenclature by Rudolf Mansfeld, Verzeichnis landwirtschaftlicher und gärtnerischer Kulturpflanzen ohne Zierpflanzen (Schultze-Motel, 1986) is used for the scientific names of the crop species (further details in Zohary et al., 2012), whereas Rothmaler, Exkursionsflora von Deutschland (Jäger et al., 2017) is used for the scientific names of the wild taxa. Several appendices give detailed information on the archaeobotanical plant assemblages: absolute find numbers per site are given in tables 1–7; new radiocarbon dates are provided in table 8. Three indices (scientific plants names, English and German plant names) list in alphabetical order those taxa that are documented on the respective plates.

The plant material presented here was exposed to a variety of mechanical, chemical, and physical impacts for thousands of years. This has resulted in differential survival of plant material and different states of preservation. This variation is documented here to indicate the limitations of taxonomic identification and to facilitate archaeobotanical investigations, by offering photographic references to fossilized charred plant remains and, in some cases, subfossil plant remains.

The variation in the shape, surface pattern, and anatomy of the surviving plant material mainly depends on the effects of depositional practices and preservation conditions. Most of the material presented here originates from sites located in areas with dry soils. Only a few of the sites are situated in the hinterland of the coast, in the immediate vicinity of or within wetland areas. While dry soil conditions hinder the preservation of organic matter because it is immediately decomposed by soil-bound micro-organisms, waterlogged environments may favour it. In dry soils with high oxygen content, plant matter survives only when it has been fully chemically transformed. This conversion of organic matter mainly takes place when it comes into contact with fire. The intensity and heat of fire, resulting in oxidizing or reductive conditions, caused plant remains to either burn to ash or survive as 'fossilized', charred remains (Jacomet and Kreuz, 1999: 59, 137). Charring experiments have shown that cereal grains undergo compaction during contact with fire (i.e. Charles et al., 2015; Hubbard and al Azm, 1990). They slightly shrink in length and become wider, approaching the cosmic perfection of a ball (Hopf, 1955). However, the effect is sufficiently minimal that visual comparison with fresh grains remains valid. The original water content of the plant remains influences the internal structure of the charred material post-charring. High water content leads to a blowy to slaggy structure, while dry and kiln-dried plant material usually shows a compact texture.

The presentation of the mainly old plant material in this volume gives an overview of Neolithic plant use in northern central Europe and shows which plant materials survive archaeologically. These are primarily the remains of crop plants, of the weed species growing among the crops on the same plots, and of those plant parts and species that were collected to supplement a diet that in Neolithic times was mainly based on intentionally grown crops. The plant economy is quite well documented in the ancient plant assemblages, because human activities favour the preservation of plant remains related to plant consumption. In particular, the use of fire in connection with cereal processing before storage or before food preparation has led to an overrepresentation of plant parts that come into contact with fire under these circumstances. Examples for such fire-related activities that further the archiving of plant residues in the ground are kiln drying of those plant staples that are stored prior to being fully processed for consumption, the cleaning of storage pits with

Fig. 1.2. Overview on the chronological settings of the well-dated archaeological sites with ancient plant assemblages presented in this volume, with the respective section numbers.

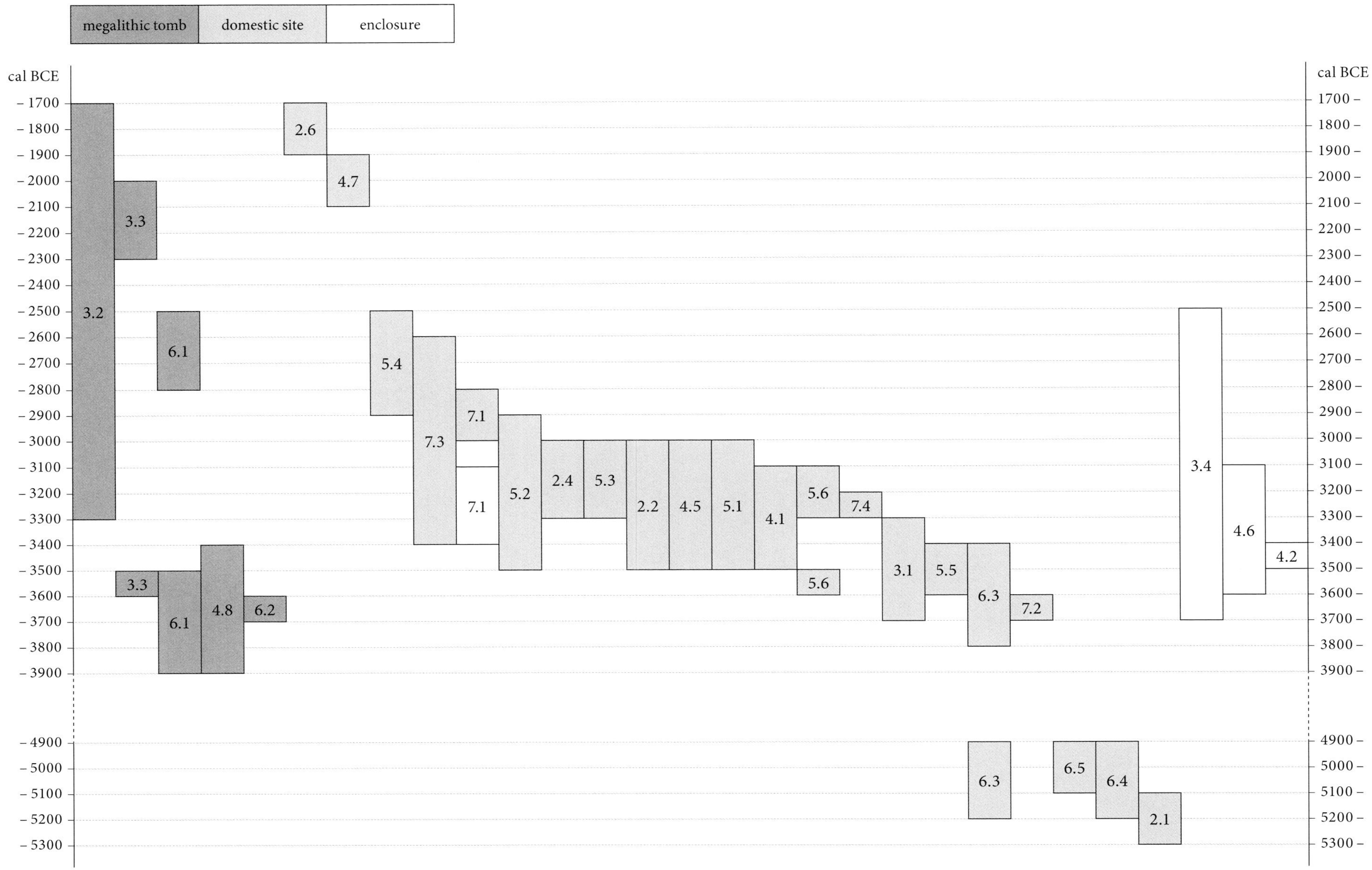

fire before re-filling, and the burning of garbage. In addition, deliberate ritual burning as well as unintentional burning of houses and farmsteads provide suitable conditions for the deposition and survival of charred plant material. The natural environment, i.e. the ruderal and the woodland flora, is usually underrepresented in the macrobotanical records from dryland archaeological sites.

Neolithic plant assemblages from central and northern Germany and from the Danish island of Funen

The plant economy in the Neolithic was mainly based on crop growing. Plant gathering complemented the crop production. The archaeobotanical material presented here relates to two different stages of the Neolithic. The first farmers in central Germany belong to the archaeological group known as the Linearbandkeramik (LBK), which dates from 5500 to 4900 BCE (Fig. 1.3). The core region of this group is congruent with the loess belt (Kreuz, 2012; Lüning, 2000). One example from this core region given here is the material from the site of Rosdorf-Mühlengrund, in southern Lower Saxony (section 2.1). However, LBK sites are known outside this main distribution area as well, beyond its northernmost border, and occupying e.g. the loamy sand to sandy loam soils of the Havelland, in Brandenburg, including the sites of Lietzow 2 and Lietzow 10 (section 6.5). The focus of this volume is mostly on a later stage of prehistory, when crop growing and animal husbandry were first established on the northern European plain, by the Funnel Beaker Culture (FBC) groups. This occurred about 1400 years after the Neolithization by LBK groups in central Europe (Gronenborn, 1999; Schier, 2009). The Funnel Beaker Culture established crop growing in the northern European plain in the period from 4100/4000 to 2800 BCE (Fig. 1.3; Kirleis et al., 2012). In this region, FBC groups were followed by those from the Neolithic Single Grave Culture (SGC) (2800 to 2200 BCE) and the Dagger Groups (2200 to 2000 BCE) (Müller, 2017, 2019). The crop spectra of the different stages of the Neolithic show characteristic compositions, indicating differing social networks or simply specific food preferences (e.g. Kirleis et al., 2012; Kirleis and Fischer, 2014; Kirleis and Willerding, 2008; Kreuz, 2012; Kreuz et al., 2014; Salavert, 2011).

The Linearbandkeramik archaeobotanical plant assemblage

The Linearbandkeramik charred plant assemblage comprises five main crop species: the cereals emmer *Triticum dicoccon* and einkorn *T. monococcum*, the legumes pea *Pisum sativum* and lentil *Lens culinaris* and the oil plant linseed *Linum usitatissimum* (Kreuz, 2012). There are some hints, from finds of stored grain, that einkorn and emmer were sometimes grown as maslins (mixed crops), sometimes even together with pea. In addition, the LBK farmers maintained separate lentil and linseed plots. Barley *Hordeum vulgare* and opium poppy *Papaver somniferum* are recorded from the Flomborn or LBK phase II onwards, representing a minor component of the crop spectrum (Kirleis and Willerding, 2008: 143; Kreuz et al., 2005). Opium poppy is a species of western Mediterranean origin. Its domestication may have taken place in multiple locations, possibly also in central Germany and in Belgium (Salavert, 2011). The gathered plants crab apple *Malus sylvestris* and hazel *Corylus avellana* supplemented the LBK plant food spectrum. The potential weed species recovered and their characteristics make intensive garden cultivation the most plausible subsistence regime for the LBK (Bogaard, 2004;

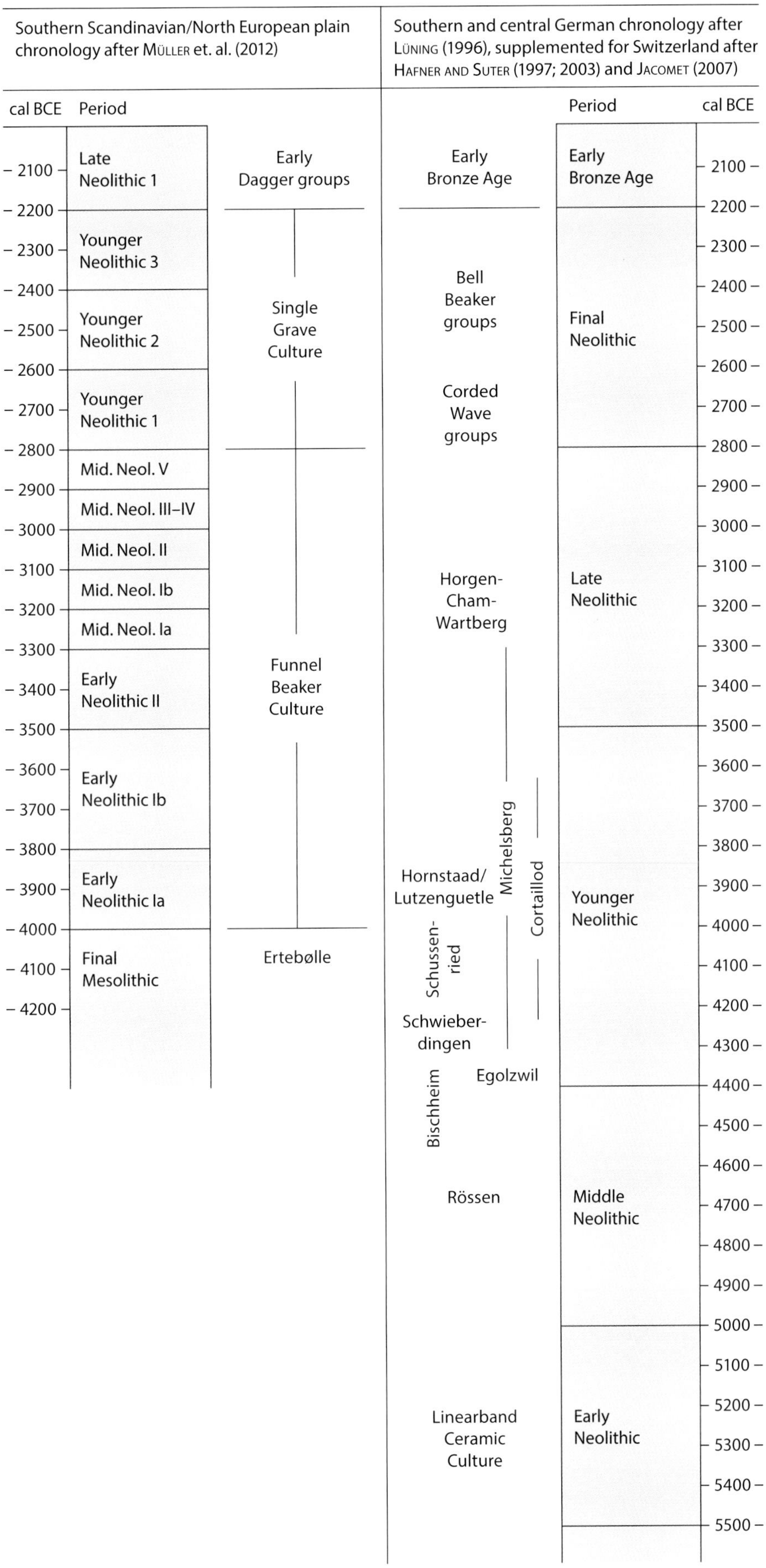

Fig. 1.3. Archaeological chronologies for the Neolithic of the southern Scandinavian/ north European plain and southern/central Germany and Switzerland (after Kirleis and Fischer, 2014: Fig. 1).

Kreuz et al., 2005: Table 7). New insights, based on a much broader diversity of Linearbandkeramik plant use present in the charred assemblages, are available from the investigations of well features with waterlogged preservation conditions (Herbig et al., 2013; Knörzer, 1998). In addition to the above-listed crop species, the western Saxonian LBK wells contained free-threshing wheat *Triticum* cf. *aestivum* s.l., the 'new type' hulled wheat cf. *Triticum timophevii,* and lentil *Lens culinaris.* Further additional potentially cultivated species are present, such as henbane *Hyoscyamus niger,* chive *Allium* cf. *schoenoprasum,* and winter cherry *Physalis alkekengi* (Herbig et al., 2013).

The Funnel Beaker archaeobotanical plant assemblage

The crop spectrum of the Funnel Beaker farmers comprises cereals and oil plants. Pulses are missing in the assemblages and presumably did not contribute to the daily nutrition of these northern farmers (Bakels, 2014). Among the cereals, the hulled wheat emmer and free-threshing barley were the preferred species. The cereal spectrum was supplemented by the hulled wheat einkorn and free-threshing wheat. Only a few sites allow for further specification of the free-threshing wheat, due to the presence of threshing remains. In the enclosure site Diekosknöll LA68 (section 3.4), a ceramic plate with imprints of rachis segments proves that the tetraploid variety of *Triticum durum/turgidum* was present in the northern Early Neolithic. The site of Frydenlund, on Funen, Denmark (section 7.2), yielded evidence of charred rachis segments, showing that the tetraploid variety of *Triticum durum/turgidum* presumably was grown in the north in the Early Neolithic. A revisiting of the archived material of the late Middle Neolithic site of Wangels LA505 (section 5.4), for the purpose of taking the photographs presented here, brought to light additional proof for the growing of free-threshing wheat with tetraploid genomic constitution, as two rachis segments of wheat could be further specified as showing the diagnostic, prominent bulbs in the rachis segments' top section, below the shoulder of the glume base insertion. All these finds are exciting, as they supply an argument for the existence of a south-west–oriented network among the Funnel Beaker groups from an archaeobotanical perspective. Free-threshing wheat with tetraploid genomic constitution entered central Europe in Neolithic times, possibly through the Rhône valley. It was a relevant component of the cereal spectrum of the Bischheim and Michelsberg archaeological groups (Kirleis and Fischer, 2014; Kreuz et al., 2014) and does not occur in the context of the LBK.

Two oil plants belong to the Funnel Beaker crop spectrum. These are linseed and opium poppy. In general, oil plants hardly contribute to the archaeobotanical record. The processing and preparation of foodstuffs made from these species does not necessarily involve fire. If seeds and fruits do come into contact with fire, they will almost invariably burn to ash because of the very high oil content. Therefore, finds of charred seeds from oil plants are extremely rare. In the north, sites offering wet and oxygen-free preservation conditions and undisturbed sediment and organic matter deposition are scarce. For this reason, finds of waterlogged seeds from oil plants are also scarce in this region. The (low) find numbers are not at all representative, and they do not inform us about the relevance of these species in the past. In these cases, it is the simple presence of the seeds that provides the evidence for the growing and use of oil plants in the past. Only two Funnel Beaker sites show evidence for the use of oil plants. These are the Early Neolithic site of Rastorf LA6a/c (section 4.1), with two charred linseeds, and the late Middle Neolithic site of Wangels LA505 (section 5.4), with waterlogged seeds of opium poppy.

Legumes are not considered to be part of the Funnel Beaker crop assemblage; emmer and barley are the crop plants commonly encountered (Bakels, 2014). The single possible pea find at Friedrichsruhe, Fpl. 8 (section 6.1), is thus suspicious for a Funnel Beaker context. Only the site of Rullstorf 8, in Lower Saxony (section 2.2), shows evidence for potential peas and seeds of vetches, but these originate from windthrow pits and thus likely date to a later archaeological period. The classical set of gathered plants resembles the LBK spectrum, as it mainly comprises the woody wild taxa, such as crab apple and hazel. Wild taxa, among them weed and herb species as well as woody taxa, accompany the crops in the assemblages. Closed contexts and finds of stored plants remains are rarely encountered, which makes it difficult to understand what species in fact grew together with the crops on the arable plots. Pollen analytical investigations stress the relevance of ribwort plantain *Plantago lanceolata* in the Funnel Beaker period (Fig. 4.1). Its charred seeds occur in low numbers in eight out of the 36 sites presented here (Tables 1, 2, 3, 4, and 6). Together with timothy grass *Phleum pratense,* it is a perennial species with a life-form as hemicryptophyte, surviving in winter with resting buds at or near ground level. The increasing occurrence of perennial species in the Funnel Beaker period may have been induced by a change in agrarian technology, namely, a reduced input of labour in the maintenance of arable plots. In short: less weeding was carried out. This change can be correlated with the stepwise establishment of the ard, which started in the northern late Early Neolithic (Mischka, 2013; Sørensen and Karg, 2014), and which, since Middle Neolithic times, was regularly being used with draught animals (Johannsen, 2006). The ard scratches the soil surface. It produces a pattern of parallel or criss-cross furrows in the ground (e. g. Fig. 4.1.2), with untouched soil surface in between – the latter being the perfect micro-area for the growth of perennial weeds. The expansion of arable fields, induced by the technological invention of the animal draught ard and an increase in population, is one of several strategies relating to the plant economy. Small, intensively worked garden-like plots in the vicinity of the farmsteads co-existed with large, extensively cultivated arable fields farther away from the villages. Plant gathering was an activity that could be perfectly combined with visiting the large, arable fields, since crab apples and hazel nuts would have been part of the forest edge vegetation at the wayside (Brozio et al., 2019).

The Younger Neolithic Single Grave archaeobotanical plant assemblage

Only a few sites in northern Germany provide information on plant use in the Younger and Late Neolithic, and this information cannot be considered representative. In general, the subsistence economy of Single Grave groups is characterized by a strong emphasis on animal husbandry, accompanied to a minor extent by cereal cultivation. At the Younger Neolithic site with special function of Wolkenwehe LA54, on the river Trave, in Schleswig-Holstein, barley was the main cereal. It was supplemented by emmer. Two single legume seeds, most likely of pea, were identified at this site, and these show that, in contrast to the Funnel Beaker groups, people of the Single Grave groups may have been acquainted with pea. But two pea seeds do not prove cultivation (Kirleis et al., 2012). In Denmark, reliable data are available for the crop assemblages attributed to the SGC (Klassen, 2005). The Single Grave groups in Denmark open up new landscapes, whereby domestic and ritual sites spread onto the western heathland areas with poor, sandy soils. The crop spectrum shows a clear preference for the growing of free-threshing barley, which dominates the assemblages. Since barley is the cereal with the widest ecological amplitude – able to cope with poor soil conditions and with being grown at high altitude – this crop preference may have been driven by the environmental setting. The analysis of food residue from a Single Grave beaker from a grave in Refshøjgaard, East Jutland, showed that barley, commonly consumed as mash, was as well used for beer production (Klassen, 2005: 6). Among the material presented here, one Younger Neolithic emmer mass find was excavated at the Danish site of Damsbo FSM 3527, on Funen island (section 7.3). This mass find of emmer grains was admixed with reams of hulled wheat glume bases as well as remains of gathered plants. Being a deposition relating to a single event, and thus not necessarily representative, this mass find nevertheless shows that the Single Grave plant economy in Denmark did not depend solely on free-threshing barley; the diversity of plants was greater. It may further hint to a regionalization of different Single Grave groups with respect to their crop preferences, or it may reflect a diachronic development, since, at the transition to the Late Neolithic, when the areas with nutrient-rich, loamy soils in eastern Jutland were settled, emmer once again increases in importance (Klassen 2005: 5).

The Late Neolithic Dagger groups archaeobotanical plant assemblage

The most striking change in Neolithic cereal cultivation is observed for the Late Neolithic Dagger groups in the north, when two newcomers occur in the archaeobotanical assemblages, namely, spelt *Triticum spelta* and hulled barley *Hordeum vulgare vulgare*, and the cultivation of free-threshing wheat is thus established (Effenberger, 2017, 2018; Brozio et al., 2019). Note that, from a methodological point of view, an identification as 'spelt' is accepted only when glume bases allow for reliable identification. A mass find of free-threshing wheat from the site of Bosau LA242 (section 4.7), although originating from a single event only, shows that this species is a relevant component of the Late Neolithic cereal spectrum. Another new aspect of the plant economy is the Late Neolithic/Early Bronze Age phenomenon of the presence of acorn (Karg and Haas, 1996; Mason, 1992). One example given here comes from the site of Vechta 10 (section 2.6). In the Late Neolithic, new supra-regional metallurgy exchange systems emerge (Klassen, 2004; Vandkilde, 2017). The cultivar newcomers visible in the archaeobotanical record add another new aspect to support the hypothesis that there were new exchange networks (Effenberger, 2017, 2018). The increasing participation in such exchange systems may be one reason for the renewed economic growth observable in the archaeological record, in which new status symbols were assigned great significance as a means of social differentiation (Brozio et al., 2019).

The main differences in the Neolithic crop assemblages

Comparison of the plant species used by the different Neolithic groups that survived as charred material in the ground shows similarities, but also some differences. The first farmers in central Europe, the Linearbandkeramik farmers of the late 6th to early 5th millennium BCE, based their cultivation on three plant groups, the cereals, the legumes, and the oil plants. The LBK cereals were dominated by the hulled wheat species emmer and einkorn. The Funnel Beaker Culture farmers of the 4th millennium BCE had different preferences. Their crop spectrum comprises only cereals and oil plants; the legumes however, are missing. Among the FBC cereals, emmer and free-threshing barley prevail, supplemented by free-threshing wheat, the latter most likely of tetraploid genomic constitution. The archaeobotanical database for the Single Grave Culture is weak. Free-threshing barley and emmer seem to be the main cereals grown. However, this picture may change as more settlement sites are identified, excavated, and archaeobotanically investigated. The crop assemblage of the Dagger Groups shows new networks that bring in spelt (a hulled wheat species) and hulled barley. The differences observed in the plant economies each have their own reasons. There is no monocausal explanation available; it is more likely a combination of multiple factors that, to a varying extent, led to the specific characteristics of these different Neolithic plant economies. Environmental and climatic conditions, the existence of supra-regional networks, as well as local traditions and culturally driven food preferences account for the differences observed in the respective crop assemblages.

Literature:

Bakels, C., 2014. The first farmers of the northwest European Plain: some remarks on their crops, crop cultivation and impact on the environment. Journal of Archaeological Science 51, 94–97.

Bogaard, A., 2004. Neolithic farming in central Europe: an archaeobotanical study of crop husbandry practices. Routledge, London.

Brozio, J.P., Müller, J., Furholt, M., Kirleis, W., Dreibrodt, S., Doerfler, W., Weinelt, M., Raese, H., Bock, A., 2019. Monuments and economies: what drove their variability in the Middle Holocene Neolithic? The Holocene (accepted).

Cappers, R.T.J., Bekker, R.M., Jans, J.E.A., 2006. Digital Seed Atlas of the Netherlands. Barkhuis Publishing and Groningen University Library, Groningen.

Cappers, R.T.J., Neef, R., Bekker, R.M., 2009. Digital Atlas of Economic Plants. Barkhuis Publishing and Groningen University Library, Groningen.

Charles, M., Forster, E., Wallace M., Jones G., 2015. "Nor ever lightning char thy grain": establishing archaeologically relevant charring conditions and their effect on glume wheat grain morphology. Science & Technology of Archaeological Research 1(1), 1–6. https://doi.org/10.1179/2054892315Y.0000000008.

Effenberger, H., 2017. The plant economy of the northern European Bronze Age—more diversity through increased trade with southern regions. Vegetation History and Archaeobotany 27(1), 65–74.

Effenberger, H., 2018. Pflanzennutzung und Ausbreitungswege von Innovationen im Pflanzenbau der nordischen Bronzezeit und angrenzender Regionen. Wachholtz, Büttelborn.

Gronenborn, D., 1999. A variation on a basic theme: the transition to farming in southern central Europe. Journal of World Prehistory 13(2), 123–210.

Hafner, A., Suter, P.J., 1997. Entwurf eines neuen Chronologie-Schemas zum Neolithikum des schweizerischen Mittellandes. Archäologisches Korrespondenzblatt 27, 549–565.

Hafner, A., Suter, P.J., 2003. Das Neolithikum in der Schweiz. Journal of Neolithic Archaeology. https://doi.org/10.12766/jna.2003.4.

Herbig, C., Maier, U., Stäuble, H., Elburg, R., 2013. „Neolithische Füllhörner". Archäobotanische Untersuchungen in fünf linienbandkeramischen Brunnen in Westsachsen, in: Carnap-Bornheim, C., Doerfler, W., Kirleis, W., Müller, J., Müller, U. (Eds.), Von Sylt bis Kastanas. Wachholtz, Neumünster, pp. 265–293.
Hopf, M. 1955. Formveränderungen von Getreidekörnern beim Verkohlen. Berichte der Deutschen Botanischen Gesellschaft 68, 191–193.
Hubbard, R., al Azm, A., 1990. Quantifying preservation and distortion in carbonized seeds; and investigating the history of frike production. Journal of Archaeological Science 17(1), 103–106.
Jacomet, S., 2007. Neolithic plant economies in the northern Alpine foreland from 5500–3500 cal BC, in: Colledge, S., Conolly, J. (Eds.), The Origins and Spread of Domestic Plants in Southwest Asia and Europe. Routledge, Walnut Creek, pp. 221–258.
Jacomet, S., Kreuz, A., 1999. Archäobotanik. Aufgaben, Methoden und Ergebnisse vegetations- und agrargeschichtlicher Forschung. Ulmer, Stuttgart.
Jäger, E.J., Müller, F., Ritz, C.M., Welk, E., Wesche, K., (Eds.), 2017. Rothmaler. Exkursionsflora von Deutschland. Gefäßpflanzen, Grundband, twelfth ed. Springer, Berlin.
Johannsen, N.N., 2013. Draught cattle and the South Scandinavian economies of the 4th millennium BC, Environmental Archaeology 11, 35–48.
Karg, S., Haas, J., 1996. Indidzien für den Gebrauch von mitteleuropäischen Eicheln als prähistorische Nahrungsressource, in: Comper, I., Hahn, J., Uerpmann, H.-P. (Eds.), Spuren der Jagd, die Jagd nach Spuren. Mo-Vince-Verlag, Tübingen, pp. 429–435.
Kirleis, W., Fischer, E., 2014. Neolithic cultivation of tetraploid free threshing wheat in Denmark and northern Germany: implications for crop diversity and societal dynamics of the Funnel Beaker Culture. Vegetation History and Archaeobotany 23, Supplement 1, 81–96.
Kirleis, W., Klooß, S., Kroll, H., Müller, J., 2012. Crop growing and gathering in the northern German Neolithic: a review supplemented by new results. Vegetation History and Archaeobotany 21(3), 221–242.
Kirleis, W., Willerding, U., 2008. Die Pflanzenreste aus der linienbandkeramischen Siedlung von Rosdorf-Mühlengrund, Ldkr. Göttingen, im südöstlichen Niedersachsen. Praehistorische Zeitschrift 83, 133–178.
Klassen, L., 2004. Jade und Kupfer, Untersuchungen zum Neolithisierungsprozess im westlichen Ostseeraum unter besonderer Berücksichtigung der Kulturentwicklung Europas 5500–3500 BC. Aarhus University Press, Aarhus.
Klassen, L., 2005. Getreide in der Einzelgrabkultur Jütlands. jungsteinSITE 2005, 1–25.
Knörzer, K.-H., 1998. Botanische Untersuchungen am bandkeramischen Brunnen von Erkelenz-Kückhoven, in: Koschik, H. (Ed.), Brunnen der Jungsteinzeit. Internationales Symposium in Erkelenz 27. bis 29. Oktober 1997. Rheinland-Verlag, Köln/Bonn, pp. 229–246.
Kreuz, A., 2012. Die Vertreibung aus dem Paradies? Archäobiologische Ergebnisse zum Frühneolithikum im westlichen Mitteleuropa. Berichte der Römisch-Germanischen Kommission 91, 23–196.
Kreuz, A., Marinova, E., Schäfer, E., Wiethold, J., 2005. A comparison of early Neolithic crop and weed assemblages from the Linearbandkeramik and the Bulgarian Neolithic cultures: differences and similarities. Vegetation History and Archaeobotany 14(4), 237–258.
Kreuz, A., Märkle, T., Marinova, E., Rösch, M., Schäfer, E., Schamuhn, S., Zerl, T., 2014. The Late Neolithic Michelsberg culture, just ramparts and ditches? A supraregional comparison of agricultural and environmental data. Praehistorische Zeitschrift 89, 72–115.
Lüning, J., 1996. Erneute Gedanken zur Benennung der neolithischen Perioden. Germania 74, 233–237.
Lüning, J., 2000. Steinzeitliche Bauern in Deutschland. Die Landwirtschaft im Neolithikum. Habelt, Bonn.
Mason, S.L.R., 1992. Acorns in human subsistence. Institute of Archaeology, University College, London.
Mischka, D., 2013. Die sozioökonomische Bedeutung von Pflugspuren im Frühneolithikum des nördlichen Mitteleuropas, in: Heske, I., Nüsse, H., Schneeweiß, J. (Eds.), Landschaft, Besiedlung und Siedlung. Archäologische Studien im nordeuropäischen Kontext. Festschrift für Karl-Heinz Willroth zu seinem 65. Geburtstag. Wachholtz, Neumünster, pp. 295–306.
Müller, J., 2017. Großsteingräber, Grabenwerke, Langhügel. Frühe Monumentalbauten Mitteleuropas. Theiss, Darmstadt.
Müller, J., 2019. Boom and bust, hierarchy and balance: from landscape to social meaning: megaliths and societies in northern central Europe, in: Müller, J., Hinz, M., Wunderlich, M. (Eds.), Megaliths, Societies, Landscapes: Early Monumentality and Social Differentiation in Neolithic Europe. Habelt, Bonn, pp. 29–74.
Müller, J., Brozio, J.-P., Demnick, D., Dibbern, H., Fritsch, B., Furholt, M., Hage, F., Hinz, M., Lorenz, L., Mischka, D., Rinne, C., 2012. Periodisierung der Trichterbechergesellschaften. Ein Arbeitsentwurf, in: Hinz, M., Müller, J. (Eds.), Siedlung, Grabenwerk, Großsteingrab. Studien zu Gesellschaft, Wirtschaft und Umwelt der Trichterbechergruppen im nördlichen Mitteleuropa. Frühe Monumentalität und soziale Differenzierung 2. Habelt, Bonn, pp. 29–33.
Neef, R., Cappers, R.T.J., Bekker, R.M., 2012 Digital Atlas of Economic Plants in Archaeology. Barkhuis Publishing and Groningen University Library, Groningen.
Salavert, A., 2011. Plant economy of the first farmers of central Belgium (Linearbandkeramik, 5200–5000 B.C.). Vegetation History and Archaeobotany 20(5), 321–332.
Schier, W., 2009. Extensiver Brandfeldbau und die Ausbreitung der neolithischen Wirtschaftsweise in Mitteleuropa und Südskandinavien am Ende der 5. Jahrtausends v. Chr. Praehistorische Zeitschrift 84, 15–43.
Schultze-Motel, J., 1986, Rudolf Mansfeld, Verzeichnis landwirtschaftlicher und gärtnerischer Kulturpflanzen ohne Zierpflanzen, second ed. Berlin. http://mansfeld.ipk-gatersleben.de
Sørensen, L., Karg, S., 2014. The expansion of agrarian societies towards the north — new evidence for agriculture during the Mesolithic/Neolithic transition in southern Scandinavia. Journal of Archaeological Science 51, 98–114.
Vandkilde, H., 2017. The Metal Hoard from Pile in Scania, Sweden. Aarhus University Press, Aarhus.
Zohary, D., Hopf, M., Weiss, E. 2012. Domestication of Plants in the Old World: The Origin and Spread of Cultivated Plants in West Asia, Europe and the Nile Walley, fourth ed. Oxford University Press, Oxford.

2 Lower Saxony: The low mountain range and loess region in the south and the lower moraine landscape in the north

The landscape of Lower Saxony consists of different geological units and highly diverse landscapes. The archaeological sites presented in this chapter are located in two different landscapes, the older moraine plain in the north and the low mountain range in the south. Recent Lower Saxonian landscapes were predominantly shaped in the Pleistocene, during the Saale glaciation c. 300 000 – 130 000 years ago. The formation of the low mountain range in the south dates back to pre-quaternary times. During the last glacial, the Weichsel (c. 115 000 – 11 500 years ago), glaciers no longer reached the areas south of the river Elbe. The ice advancing from the north-east into central Europe had already filled up the terminal moraines of the Weichsel glaciation in present-day Schleswig-Holstein. Thus, depositions from the previous Saale glaciation dominate in northern Lower Saxony, where the lower moraine landscape covers most parts of the flat, but undulating moraine plates. The relief was transformed during the Weichsel glaciation by periglacial processes. The formerly knobby topography closer to the terminal moraines had already been levelled out by soil flow and water erosion. The sediments of the lower moraine comprise boulder clay and marly till with differing sand, clay, and lime content. The lower moraine landscape in northern Lower Saxony extends from the river Ems, at the border to the Netherlands, to the river Elbe, in the east, being interrupted by the large river valley of the Weser and smaller tributaries to the main water courses. In the Holocene, glacio-fluviatile sands deposited e.g. in the Elbe glacial valley were windblown into dunes, which migrated eastwards under the prevailing westerly winds. Cavities created during the glacial periods by melting ice blocks or meltwater in the lower moraine plain developed from lakes into peat bogs during the Holocene and contributed to further structuring of the landscape (cf. Benda, 1995). The areas directly north of the low mountain range belong to the central European loess zone. Loess as a periglacial deposition was formed when silt and fine sand were blown out in areas with open vegetation in front of the ice sheet during the Weichsel glaciation and settled again, after prolonged transport of up to several tens to several hundreds of kilometres, in areas with denser vegetation cover that trap the fine material. After deposition, the individual minerals of the easily mobilized lime or clay content usually became cemented together to build the loess formation that prepares the ground for the development of fertile soils that are of great interest to farming communities.

In southern Lower Saxony, the distribution of settlements of the earliest central European farmers, of the Linearbandkeramik (LBK) Culture, reaches its north-western extent periphery, with 141 settlements concentrating in the loess areas and the hilly country. In the second half of the sixth millennium BCE, the main settlements were located in the south-east, in the Leine valley, between Göttingen (section 2.1) and Einbeck; on the fertile soils in the Untereichsfeld, around Gieboldehausen, Duderstadt, and Lake Seeburg; in the south-western Harz foreland; and in the area of the Helmstedter-Wolfenbütteler, Hildesheimer, and Braunschweiger Börde. LBK settlements line the valley of the river Leine in high density and link the area with LBK settlement clusters farther to the south, stretching from Hessia to Transdanubia (Saile, 2007). The distribution of LBK settlement sites in northern Germany extended only slightly beyond the Pleistocene and Late Glacial loess deposits to the north although recently, an LBK settlement was identified on the lower terrace of the river Leine, north of Hanover (Gerken and Nelson, 2016), and the Bullenteich pollen diagram, from the Braunschweig urban area, indicates another settlement of the LBK north of the loess border (Lagies, 2003: 313).

The lower moraine plain in Lower Saxony was barely populated for a long time. Small groups of Mesolithic hunter-gatherer-fishers prevailed, and these favoured the diverse coastal and lake landscapes with extended woodland borders, which, together, offered access to multiple natural resources. Farming activity in the area only began in the fourth millennium BCE, with the Swifterbant Group, and continued with Funnel Beaker Culture (FBC) groups, both of which inhabited coastal as well as interior landscapes. Most archaeological finds, such as pottery and stone artefacts, were found in the sandy moraines and the areas nowadays covered by bog. In the north-west, the bog expansion, triggered by more pronounced oceanic climatic conditions from 6000 BCE onwards, furthered the fragmentation of the sandy moraines into small islands. These moraine islands were inhabited by farming communities of the Funnel Beaker Culture from c. 4000 BCE onwards. Megalithic tombs are the dominant monuments that indicate the distribution of the Funnel Beaker Culture in northern Lower Saxony. Neolithic settlement clusters have been found e.g. in the Flögeln area, with the sites Flögeln-Eekhöltjen and Sievern 114 (section 2.5). Important insights into the development of the cultural landscape and vegetation history were inferred from archaeobotanical investigations on the Flögeln moraine island (Behre and Kučan, 1994), namely, a first onset of agriculture with woodland pasture, an increased woodland opening, and a gradual transition to full agrarian subsistence based on crop cultivation (Kramer et al., 2012). Apart from Flögeln, the sites of Lavenstedt 178 (section 2.4) and Rullstorf 8 (section 2.2), in the district of Lüneburg, are two of the few Funnel Beaker sites with remains of houses (Mennenga, 2017). The Funnel Beaker site Hittfeld 61 (section 2.3) revealed domestic contexts in the form of pits and a possible sunken floor. Domestic structures from later Neolithic periods are extremely rare. Rescue excavations at Vechta 10 (section 2.6), surprisingly, yielded as the only feature of Late Neolithic origin a mass find of barley grains.

Literature:

Behre, K.-E., Kučan, D., 1994. Die Geschichte der Kulturlandschaft und des Ackerbaus in der Siedlungskammer Flögeln, Niedersachsen, seit der Jungsteinzeit. Probleme der Küstenforschung im südlichen Nordseegebiet 21, August Lax, Hildesheim.

Benda, L., 1995. Das Quartär Deutschlands. Gebrüder Bornträger, Berlin, Stuttgart.

Gerken, K., Nelson, H., 2016. Niederstöcken 21. Linienbandkeramisches Expansionsgebiet jenseits der Lössgrenze im Land der Jäger und Sammler? Nachrichten aus Niedersachsens Urgeschichte 85, 31–84.

Kramer, A., Mennenga, M., Nösler, D., Jöns, H., Bittmann, F., 2012. Neolithic Settlement and Land Use History in Northwestern Germany—First Results from an Interdisciplinary Research Project, in: Hinz, M., Müller, J. (Eds.), Siedlung, Grabenwerk, Großsteingrab. Studien zu Gesellschaft, Wirtschaft und Umwelt der Trichterbechergruppen im nördlichen Mitteleuropa. Habelt, Bonn, pp. 317–336.

Lagies, M., 2003. Untersuchungen zur nacheiszeitlichen Vegetationsgeschichte am Bullenteich in Braunschweig. Neue Ausgrabungen und Forschungen in Niedersachsen 24, 297–324.

Mennenga, M., 2017. Zwischen Elbe und Ems. Die Siedlungen der Trichterbecherkultur in Nordwestdeutschland. Habelt, Bonn.

Saile, T., 2007. Aspekte der Neolithisierung in der Weser-Harz-Region. Germania 85, 181–197.

2.1 Rosdorf-Mühlengrund

Site name	Rosdorf-Mühlengrund
Longitude, latitude	51.503614, 9.903106
Federal state, country	Lower Saxony, Germany
Excavation campaign	1963–1970
Excavation type	Research excavation
Responsible archaeologist	Brigitte Schlüter, Göttingen University
Archaeological context	Settlement site
Analysed structures	Pits, postholes
Archaeological group	Linearbandkeramik
Archaeological periodisation	Early Neolithic (central Europe)
Absolute dating	c. 5300–5150 cal BCE
Archaeobotanists	Ulrich Willerding, Wiebke Kirleis, Göttingen
Archaeobotanical sampling strategy	Judgement sampling from archaeological contexts
Analysed sediment volume	> 2000 l
Find concentration (finds/l)	14
Number of carpological remains	28685

The LBK site of Rosdorf-Mühlengrund, in the southwest of Göttingen is situated in the Weser-Harz region of southern Lower Saxony, in what was part of the densely populated Middle Elbe region in the second half of the sixth millennium BCE. In central Europe, the earliest farmers to carry out small-scale crop cultivation belong to the LBK. The crop spectrum of LBK farmers comprised mainly the hulled wheats emmer and einkorn. The pulses lentil and pea as well as the oil plant linseed supplemented the cereals. Within a short time, all these crops were further supplemented by barley and opium poppy. Rosdorf-Mühlengrund is the LBK site with the largest excavated area in southern Lower Saxony, with 36 trenches excavated over an area of 23 000 m^2. Altogether, 52 LBK house plans were discovered (Schlüter, 1983). The excavations were exceptional because, from the very beginning, in the 1960s, natural scientific investigations were taken into consideration. Thus, large volumes of sediment were floated through fine mesh sieves to recover botanical and zoological remains.

The archaeobotanical assemblage (Table 1) comprises, in addition to cereals, caryopses and seeds of pulses, the tiny glume bases and spikelet forks of emmer and einkorn, as well as seeds and fruits of weeds typical for the Bromo-Lapsanetum praehistoricum (Knörzer, 1971). Particularly noteworthy is that here, for the first time, material from a single site in southern Lower Saxony represents three phases of the Linearbandkeramik. The biggest amount of plant remains originates from the transition from the early to early-to-middle LBK, from house XVII (Fig. 2.1.1). There is considerable evidence from the early-to-middle phase, also known as the Flomborn Phase, but the find of an opium poppy seed hints at a late LBK date for the ground plan of this house (Kirleis and Willerding, 2008; Willerding, 1965). This chronological sequence makes it possible to demonstrate settlement-specific changes in the range of cultivated plants that were grown, on permanent plots, in the direct vicinity of the site.

Fig. 2.1.1. Distribution of charred plant remains at the transition from the early to the early-to-middle Linearbandkeramik in the ground floor of house XVII, Rosdorf-Mühlengrund (after Kirleis and Willerding, 2008: Fig. 1).

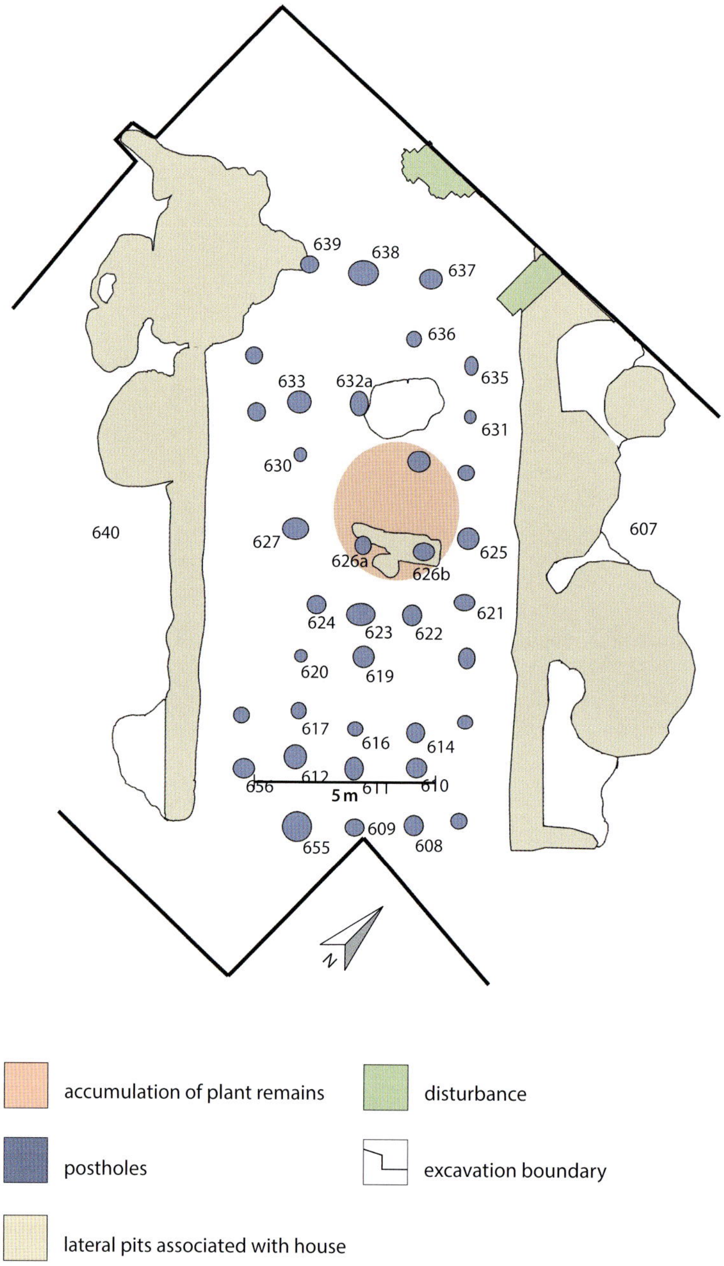

Literature:

Kirleis, W., Willerding, U., 2008. Die Pflanzenreste der linienbandkeramischen Siedlung von Rosdorf-Mühlengrund, Ldkr. Göttingen, im südöstlichen Niedersachsen. Prähistorische Zeitschrift 83(2), 133–178.

Knörzer, K.-H., 1971. Urgeschichtliche Unkräuter im Rheinland. Ein Beitrag zur Entstehung der Segetalgesellschaften. Vegetatio, 23(1–2), 89–111.

Schlüter, B., 1983. Die bandkeramische Siedlung auf dem Mühlengrund in Rosdorf, Ldkrs. Göttingen, in: Wegener, G. (Ed.), Frühe Bauernkulturen in Niedersachsen. Archäologische Mitteilungen aus Nordwestdeutschland. Beiheft 1. Isensee Verlag, Oldenburg, pp. 45–90.

Willerding, U., 1965. Die Pflanzenreste aus der bandkeramischen Siedlung. Neue Ausgrabungen und Forschungen in Niedersachsen 2, 44–60.

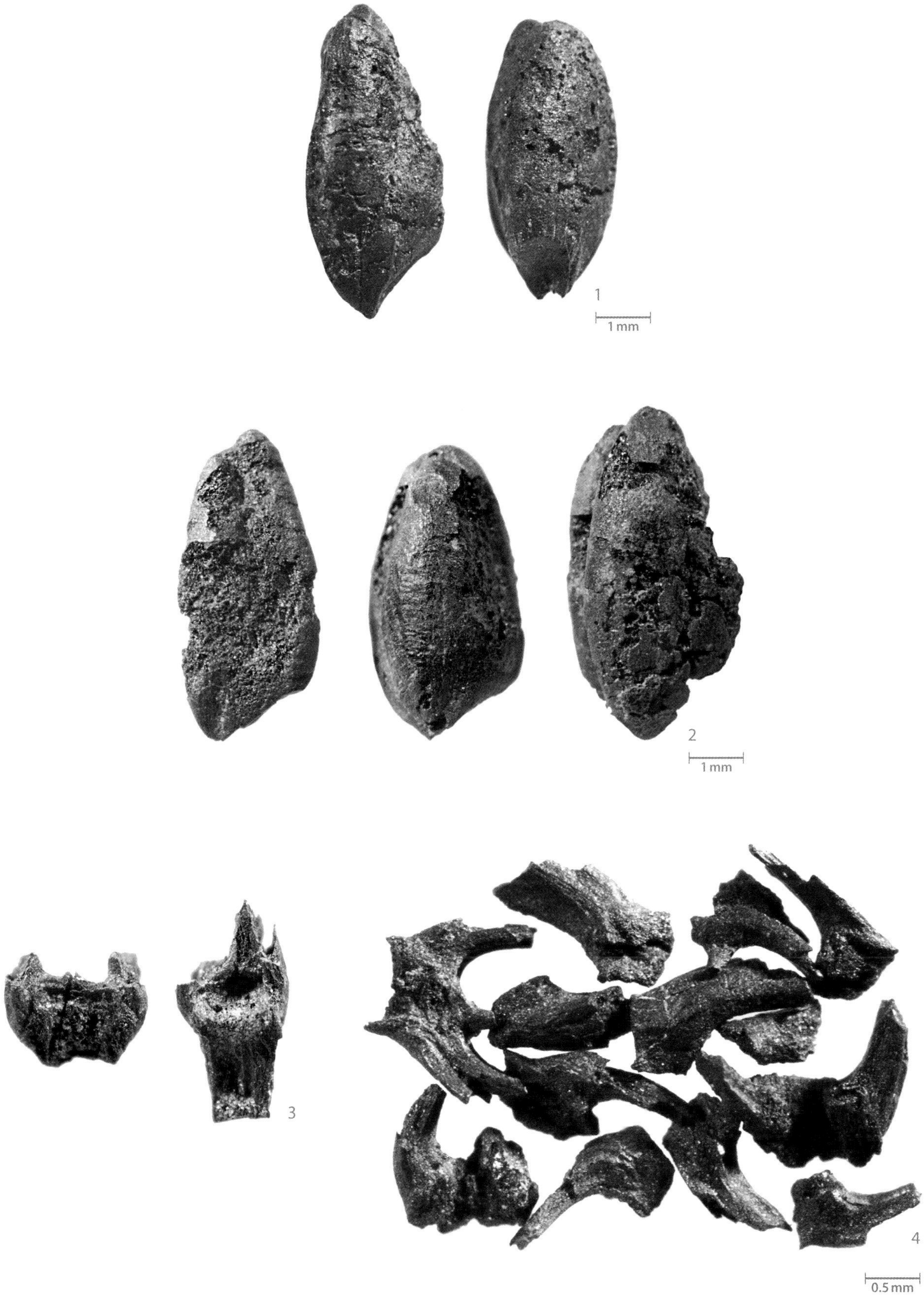

Plate 1. Rosdorf-Mühlengrund. 1 and 2 Charred einkorn grains, *Triticum monococcum*, lateral and dorsal views. 3 Charred spikelet forks of einkorn. 4 Charred spikelet forks of emmer/einkorn, *T. dicoccon/monococcum*.

Plate 2. Rosdorf-Mühlengrund. 1 and 2 Charred emmer grains, *Triticum dicoccon*, dorsal and lateral views. 3 Charred glume bases of emmer.

Plate 3. Rosdorf-Mühlengrund. 1 Charred grain of barley, *Hordeum vulgare*, ventral view. 2 and 3 Charred linseeds, *Linum usitatissimum*. 4 Charred hilum of pea, *Pisum sativum*. 5 Charred peas. 6 Charred nutshell fragment of hazel, *Corylus avellana*. 7 Charred fragment of a caryopsis of brome, *Bromus* sp. 8 Charred fragments of black bindweed pericarp, *Fallopia convolvulus*.

Plate 4. Rosdorf-Mühlengrund. 1 Charred seed of white goosefoot, *Chenopodium album*. 2 Charred seed of cleavers, *Galium aparine*, outer view. 3 Charred seed of false cleavers, *Galium spurium*, outer view. 4 Charred seed of black nightshade, *Solanum nigrum*. 5 Charred caryopses of barnyard grass, *Echinochloa crus-galli*. 6 Charred caryopsis of annual meadow grass, *Poa annua*. 7 Charred fruit of common woodrush, *Luzula luzuloides*. 8 Charred seed fragment of common nipplewort, *Lapsana communis*.

2.2 Rullstorf 8

Site name	Rullstorf 8
Longitude, latitude	53.285789, 10.530626
Federal state, country	Lower Saxony, Germany
Excavation campaign	1979–2009
Excavation type	Rescue excavation
Responsible archaeologist	Wilhelm Gebers, Niedersächsisches Landesamt für Denkmalpflege
Archaeological context	Settlement site
Analysed structures	Pits, postholes, cultural layer, pits resulting from tree windthrow
Archaeological group	Tiefstichkeramik group of the Funnel Beaker Culture
Archaeological periodisation	Early Neolithic to Middle Neolithic (EN II–MN II)
Absolute dating	c. 3500–3050 cal BCE
Archaeobotanists	Stefanie Klooß, Wiebke Kirleis, Kiel (SPP1400)
Archaeobotanical sampling strategy	Judgement sampling from archaeological contexts
Analysed sediment volume	c. 1640 l
Find concentrations (finds/l)	1 (0.3)
Number of carpological remains	477

The multi-period site of Rullstorf is situated in north-eastern Lower Saxony, district Lüneburg, on the Kronsberg lower moraine spur, facing the river Elbe (Kirleis, 2003). The favourable site was recurrently re-visited in prehistoric and historic times, starting with the Funnel Beaker settlement and followed by an urnfield graveyard of 130 burials associated with a late Bronze Age settlement site with numerous storage pits; a pre-Roman Iron Age phase with house structures, domestic pits, and kilns and another urnfield graveyard; long houses and smelting pits for iron ore processing from the Roman Iron Age to the migration period; and a late Saxon elite burial with a pack of hounds and deer (Lauxtermann et al., 2003). The oldest part of the site, Rullstorf 8, dates to the Funnel Beaker period. Six new radiocarbon dates on charred material from the domestic site calibrate to 3500–3050 cal BCE, thus the Early Neolithic II to Middle Neolithic I/II (Table 8). A pit and 17 postholes allow for the reconstruction of a two-aisled house measuring 24 × 5 m, for a large overall surface area of 122 m² (Fig. 2.2.1; Mennenga, 2017). Reams of ceramic sherds and stone artefacts were recovered. Postholes and pits are the most important features, but the content of windthrow pits, particularly in the house area, show additional evidence for the Funnel Beaker Culture and thus are considered here. The ceramics are characteristic of the Tiefstich group of the Funnel Beaker Culture, which occurs north of the Harz mountains.

The Neolithic archaeobotanical assemblage (Table 1) includes cereal grains and chaff remains of emmer and einkorn and a single grain of barley. Emmer

Fig. 2.2.1. Funnel Beaker house plan, Rullstorf 8 (after Mennenga, 2017: Fig. 45).

and einkorn are evidenced predominantly by their glume bases. Exceptional for the Funnel Beaker Culture is the evidence for legumes as further potential crop plants, but since these seeds were found in windthrow pits, they may be intrusive. If not, the Tiefstich group may have had specific food preferences. Gathered plants comprise hazel, raspberry, bramble, wild strawberry, and hawthorn, as well as the starch-rich tubers of lesser celandine. The common Neolithic weed spectrum with cleavers, knotweed, ribwort plantain, brome, and timothy/annual meadow grass is supplemented by bugleweed *Ajuga* sp. A large proportion of the postholes contained no charred plant remains at all; the majority of the glume bases were found in just seven postholes, accompanied by fruits of Poaceae and seeds of ribwort plantain. Two features archaeologically described as pits resulting from tree windthrow that contained Neolithic material have peculiar plant assemblages (Features 5422 and 5282). In addition to cereal remains and different weed seeds, they yielded a large number of charred vetch seeds and cone scales of pine.

Literature:

Kirleis, W., 2003. Vegetationsgeschichtliche und archäobotanische Untersuchungen zur Landwirtschaft und Umwelt im Bereich der prähistorischen Siedlungen bei Rullstorf, Ldkr. Lüneburg. Probleme der Küstenforschung im südlichen Nordseegebiet 28, 65–132.

Lauxtermann, B., Gebers, W., Moser, A., Wachendorf, P., 2003. Von der Jungsteinzeit bis zum frühen Mittelalter. Eine vorläufige Bilanz der letzten Grabungen in Rullstorf bei Lüneburg. Berichte zur Denkmalpflege in Niedersachsen 23(1), 3–11.

Mennenga, M., 2017. Zwischen Elbe und Ems. Die Siedlungen der Trichterbecherkultur in Nordwestdeutschland. Habelt, Bonn.

Plate 1. Rullstorf 8. 1 Charred einkorn grain, *Triticum monococcum*, lateral view. 2 Charred glume base of einkorn. 3 Charred spikelet fork of emmer, *Triticum dicoccon*. 4 Charred glume bases of emmer. 5 Charred seeds of cultivated legumes, Leguminosae sativae, cf. *Pisum*.

Plate 2. Rullstorf 8. 1 Charred nutshell fragments of hazel, *Corylus avellana*. 2 Charred seed of bugleweed, *Ajuga* sp., dorsal and ventral view. 3 Charred fruits of black bindweed, *Fallopia convolvulus*. 4 Charred seed of ribwort plantain, *Plantago lanceolata*, lateral view.

Plate 3. Rullstorf 8. 1 Charred seeds of bramble, *Rubus fruticosus*. 2 Charred seeds of raspberry, *Rubus idaeus*. 3 Charred nutlet of wild strawberry, *Fragaria vesca*. 4 Mass find of charred seeds of vetches, *Vicia* sp.

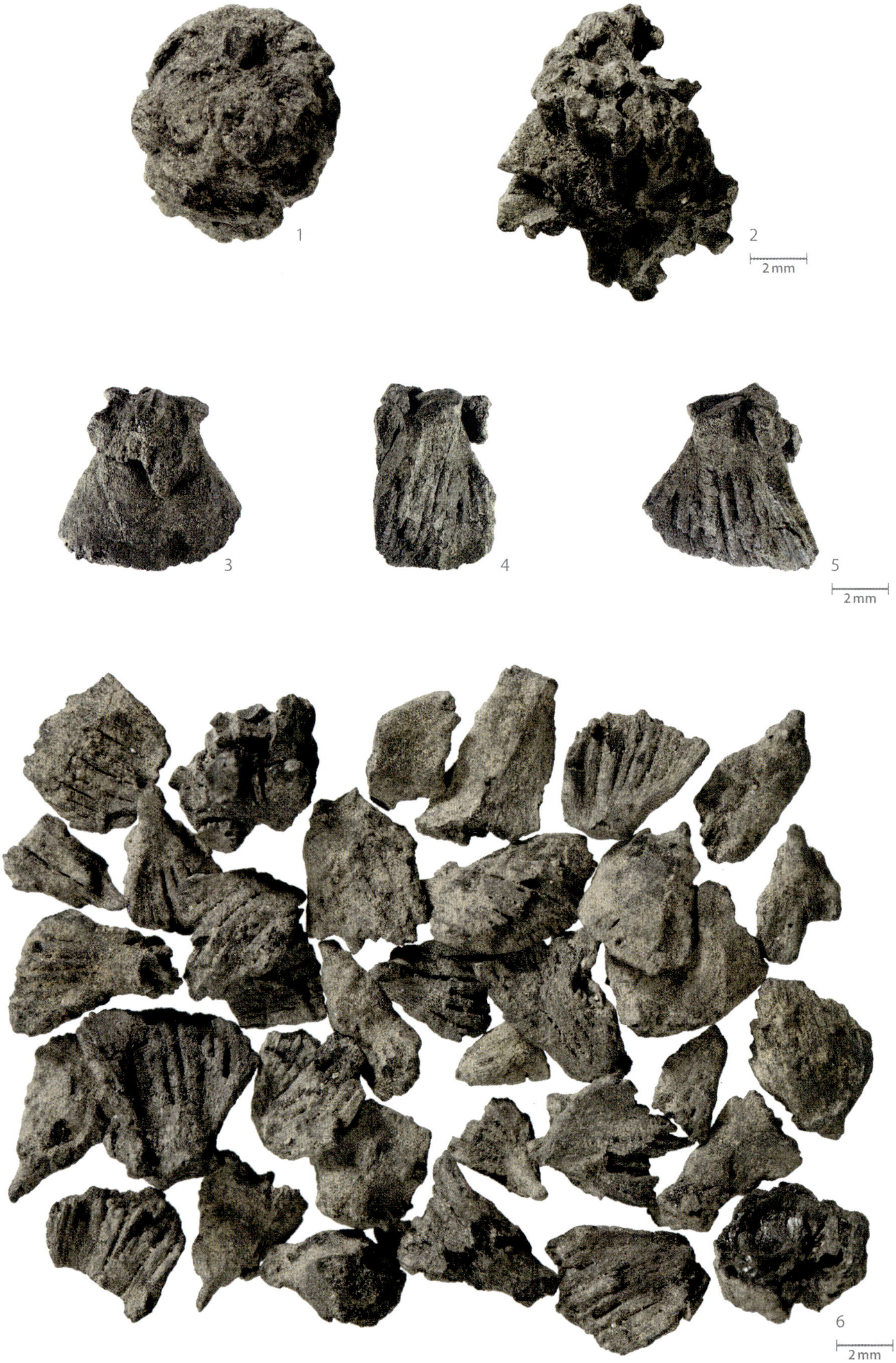

Plate 4. Rullstorf 8. 1 and 2 Charred tiny cone of pine, *Pinus sylvestris*. 3 – 5 Charred cone scale of pine, outer, lateral, and inner views. 6 Charred cone scales of pine.

2.3 Hittfeld 61

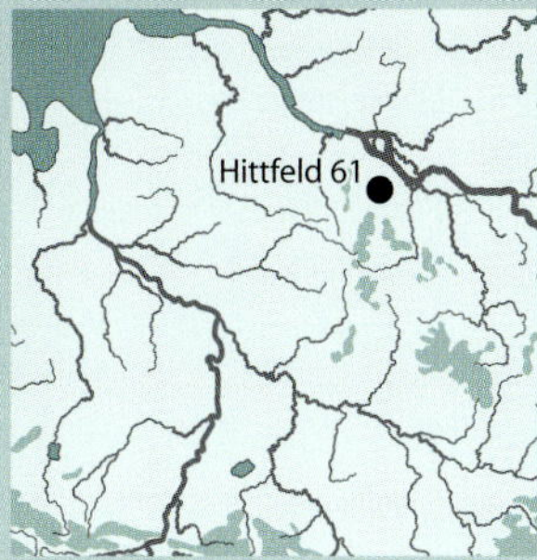

Site name	Hittfeld 61
Longitude, latitude	53.392316, 10.004316
Federal state, country	Lower Saxony, Germany
Excavation campaign	2014
Excavation type	Rescue excavation
Responsible archaeologist	Archaeological Museum Hamburg
Archaeological context	Settlement site
Analysed structures	Pits, postholes
Archaeological group	Funnel Beaker Culture
Archaeological periodisation	–
Absolute dating	–
Archaeobotanists	Stefanie Klooß, Wiebke Kirleis, Kiel (SPP1400)
Archaeobotanical sampling strategy	Judgement sampling from archaeological contexts
Analysed sediment volume	c. 281 l
Find concentrations (finds/l)	1 (0.8)
Number of carpological remains	219

The site of Hittfeld 61 was excavated prior to the establishment of a cable route close to Hittfeld, in the district of Harburg. The excavation revealed features from the Neolithic, mainly the Funnel Beaker Culture, and one pit with an arrowhead typical for the Single Grave Culture. Further pits date to the Roman Iron Age. A possible sunken floor with an aggregate of pits contains numerous Funnel Beaker artefacts, as well as charred plant material: ceramic sherds; silex flakes, scrapers, micro-blades; a flat axe; and an almost complete shouldered vessel with four handles and line decoration on the shoulder that had been stuck upside down in the ground (Fig. 2.3.1, 2.3.2; Tröger, 2016).

The archaeobotanical material (Table 1) from the Funnel Beaker Culture stems from 25 features. Most of the remains, including all 50 emmer grains;

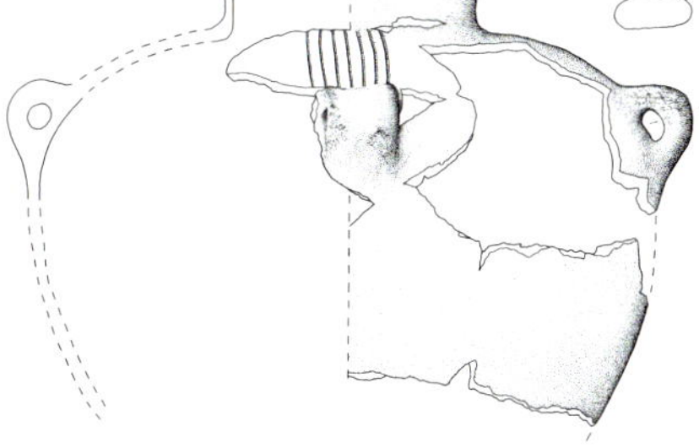

Fig. 2.3.1. Funnel Beaker amphora positioned upside down in the ground, Hittfeld 61 (photo: Freia Tröger, ArchaeoFirm, Isernhagen, for the Archaeological Museum Hamburg).

Fig. 2.3.2. Funnel Beaker amphora shown in Fig. 2.3.1, Hittfeld 61 (drawing: Nina Batram for the Archaeological Museum Hamburg).

Fig. 2.3.3. Rectangular feature representing a sunken floor, with numerous Funnel Beaker and botanical artefacts, Hittfeld 61 (photo: Freia Tröger, ArchaeoFirm, Isernhagen, for the Archaeological Museum Hamburg).

29 emmer glume bases; some barley; and 140 further cereal grains, bramble seeds, and crab apple remains, were found in the sunken floor area (Fig. 2.3.3). Weed remains are rare. The activity area with the sunken floor thus gives insight into the material culture and the culinary preferences of the Funnel Beaker Culture. Hittfeld 61 is one example for the occurrence of charred lime fruits, often unripe, in domestic and ritual sites of the Funnel Beaker Culture. The almost regular occurrence of charred lime (linden) fruits has multiple reasons. Lime was a common tree in the Neolithic mixed oak forests in northern Germany before beech started to expand in this region. Lime trees show an extremely high fruit production. Further, lime has multiple uses, ranging from use of the light timber, e.g. for dug-out canoes; to use of the flowers as medical plant; to use of the bast for fibre production.

Literature:

Tröger, F., 2016. 123 Hittfeld Fdst. Nr. 61, Gde. Seevetal, Ldkr. Harburg. Fundchronik Niedersachsen 2014, NNU Beiheft 19, 89–91.

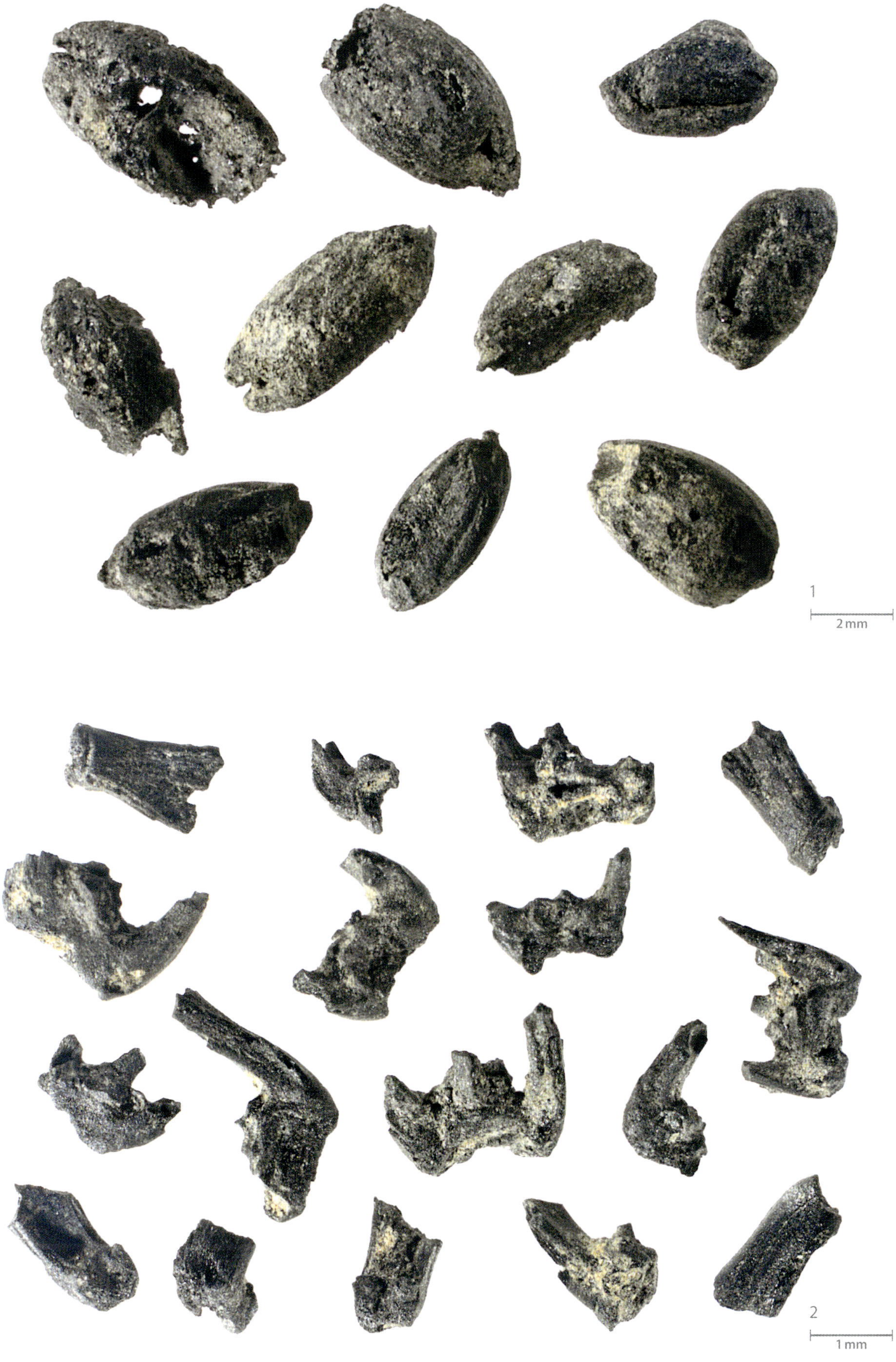

Plate 1. Hittfeld 61. 1 Charred emmer grains, *Triticum dicoccon*. 2 Charred spikelet forks and glume bases of emmer.

Plate 2. Hittfeld 61. 1 – 3 Charred barley grains, *Hordeum vulgare*, dorsal and ventral views. 4 – 6 Charred skin and pulp of crab apple, *Malus sylvestris*. 7 Charred seed of bramble, *Rubus fruticosus*. 8 Charred seed of cleavers, *Galium aparine*. 9 Charred seed of lime, *Tilia* sp., outer and interior views.

2.4 Lavenstedt 178

Site name	Lavenstedt 178
Longitude, latitude	53.336426, 9.202496
Federal state, country	Lower Saxony, Germany
Excavation campaign	2009–2010; 2011–2013
Excavation type	Research excavation
Responsible archaeologists	Klaus Gerken, Stefan Hesse, Kreisarchäologie Rotenburg/Wümme; Daniel Nösler, Moritz Mennenga, Hauke Jöns, NIhK Wilhelmshaven
Archaeological context	Settlement site
Analysed structures	Cultural layer, pits, postholes
Archaeological group	Funnel Beaker West group
Archaeological periodisation	Middle Neolithic (MN I–MN II; Brindley 3–5)
Absolute dating	c. 3350–3000 cal BCE
Archaeobotanists	Stefanie Klooß, Wiebke Kirleis, Kiel (SPP1400)
Archaeobotanical sampling strategy	Systematic grid-wise sampling (every fifth quadrant) in the cultural layer plus judgement sampling from archaeological contexts
Analysed sediment volume	c. 6019 l
Find concentrations (finds/l)	1 (0.2)
Number of carpological remains	990

The Neolithic settlement site of Lavenstedt 178 is situated on a spur of the Zeven lower moraine ridge, facing the lowlands of the river Oste, in northern Lower Saxony. Thus, the site is located partly under forest cover and partly in a wet meadow. Several springs exist in the direct vicinity (Kramer et al., 2012). Archaeologically, Lavenstedt 178 is characterized by a 20 to 40 cm thick cultural layer with a re-worked hard pan horizon that is full of artefacts (Fig. 2.4.1). Preliminary investigations on a limited area of 60 m^2 yielded about 1000 kg of finds, namely, rocks, flint artefacts, ceramics, and three amber beads. Features below the cultural layer tend to be diffuse and shallow, and they are poorly visible because of leaching and erosion processes in the soil. However, three find complexes with postholes could be interpreted as buildings, namely, two houses and a smaller structure with numerous lithic artefacts (Fig.2.4.2; Mennenga, 2017: 99). Beside others, concentrations of grinding stones and a possible well indicate domestic activity at the site.

The archaeobotanical material from Lavenstedt 178 is poorly preserved and concentrations are low (Table 1). Most of the charred finds originate from the well (Feature 283) and mainly comprise indeterminate cereals (Cerealia indet.). Identification to the species level was rarely possible. About 68 per cent of the cereal grains are fragmented and blowy to slaggy (Plate 1.1). The remaining 32 per cent are dominated by free-threshing barley. Emmer is second in importance among the cultivated species, and other cereals are

Fig. 2.4.1. Soil profile showing dark cultural layer, a cut hard pan horizon above in situ meltwater sand, Lavenstedt 178 (Mennenga, 2017: Fig. 29).

present in minor proportions, with only single grains of free-threshing wheat and a single spikelet fork of einkorn indicating the presence of these species. The robust nutshells of hazel survived, clustering in particular around a feature with stone paving of unknown function. Single finds of crab apple, bramble, and rose complement the assemblage of gathered plants. The weeds comprise common species, such as knotweed species, white goosefoot, ribwort plantain and clover, as well as tiny caryopses from the grass family, possibly *Poa annua* and *Phleum pratense*.

Fig. 2.4.2. Lavenstedt 178. Stone pavement 1 (photo: Mennenga, 2017: Fig. 68).

Literature:

Kramer, A., Mennenga, M., Nösler, D., Jöns, H., Bittmann, F., 2012. Neolithic Settlement and Land Use History in Northwestern Germany—First Results from an Interdisciplinary Research Project, in: Hinz, M., Müller, J. (Eds.), Siedlung, Grabenwerk, Großsteingrab. Studien zu Gesellschaft, Wirtschaft und Umwelt der Trichterbechergruppen im nördlichen Mitteleuropa. Habelt, Bonn, pp. 317–336.

Mennenga, M., 2017. Zwischen Elbe und Ems. Die Siedlungen der Trichterbecherkultur in Nordwestdeutschland. Habelt, Bonn.

Plate 1. Lavenstedt 178. 1 Charred blowy-slaggy cereal grains, Cerealia indet. and a variety of seeds of the knotweed family, Polygonaceae. 2 – 4 Charred grains of free-threshing barley, *Hordeum vulgare nudum*, ventral view. 5 Charred puny grain of barley, *Hordeum vulgare*, ventral view.

Plate 2. Lavenstedt 178. 1 Charred spikelet fork of einkorn, *Triticum monococcum*. 2 Charred tiny caryopses of the grass family, Poaceae. 3 Charred seed of white goosefoot, *Chenopodium album*. 4 Charred remains of black bindweed, *Fallopia convolvulus*.

2.5 Sievern 114

Site name	Sievern 114
Longitude, latitude	53.649083, 8.605534
Federal state, country	Lower Saxony, Germany
Excavation campaign	2010–2011
Excavation type	Research excavation
Responsible archaeologists	Daniel Nösler, Hauke Jöns, NIhK Wilhelmshaven
Archaeological context	Settlement site
Analysed structures	Pits, postholes
Archaeological groups	Funnel Beaker West group; Single Grave Culture
Archaeological periodisation	Neolithic
Absolute dating	–
Archaeobotanists	Stefanie Klooß, Wiebke Kirleis, Kiel (SPP1400)
Archaeobotanical sampling strategy	Judgement sampling from archaeological contexts
Analysed sediment volume	c. 106 l
Find concentrations (finds/l)	1 (0.1)
Number of carpological remains	7

The site of Sievern 114 is situated in the lower moraine micro-region of Flögeln/Sievern, in north-western Lower Saxony. It is a landscape characterized by lower moraine summits interspersed with wet depressions, where peat growth started around 6000 BCE, triggered by a transgression of the North Sea that resulted in pronounced oceanic climatic conditions (Behre, 2005). The vegetation and settlement history of the area has been the subject of intensive interdisciplinary study since the 1970s, bringing together archaeologists and natural scientists and including macro-botanical remains from the excavations and pollen from natural archives, namely, 12 kettle hole bogs (Behre and Kučan, 1994). The results constitute one of the first comprehensive studies in landscape archaeology in northern Germany with a focus on the Neolithic period. Early human activity in the area is indicated by the presence of megalithic tombs. Detailed archaeological excavations have revealed Neolithic settlement structures and allowed for reconstruction of Funnel Beaker house floor plans. The respective pollen records in combination with the on-site studies explain the stepwise development of a fully agrarian Neolithic. Re-investigations in the area have provided revised chronologies for some of the pollen profiles but support the observation of a gradual onset of the Neolithic, with slight woodland opening dated to to around 4200 cal BCE, and an increased openness in the regional vegetation after 3500 cal BCE that is concurrent with the archaeological evidence for Funnel Beaker activity (Kramer and Bittmann, 2015).

Fig. 2.5.1. Sievern 114: Potential posthole, feature 100 (after Mennenga 2017, fig. 36, p. 75).

A small-scale excavation was carried out in 2010/11 at the site Sievern 114. Although the features were in a poor state of preservation, they could be interpreted as pits and postholes (Fig. 2.5.1). The few artefacts recovered, such as grinding stones and flint axes, some of them no longer in archaeological context, allow for a careful interpretation as a domestic site. The picture from radiocarbon dates remains vague since they are scattered from the Early Neolithic to the Bronze Age and were measured at the Leibniz laboratory in Kiel at a time when sample processing was problematic (Mennenga, 2017: 50). The archaeobotanical assemblage is small, and it does not indicate a domestic character for the site (Table 1). Crop plants are absent. Fragments of charred nutshells of hazel indicate the gathering of plants. Further finds comprise charred scales from pine cones.

Literature:

Behre, K.-E., 2005. Die Einengung des neolithischen Siedlungsraumes in Nordwestdeutschland durch klimabedingte Faktoren: Meeresspiegelanstieg und großflächige Ausbreitung von Mooren, in: Gronenborn, D. (Ed.), Klimaveränderung und Kulturwandel in neolithischen Gesellschaften Mitteleuropas, 6700 – 2200 v. Chr. RGZM-Tagungen 1, Römisch-Germanisches Zentralmuseum, Mainz, pp. 209 – 220.

Behre, K.-E., Kučan, D., 1994. Die Geschichte der Kulturlandschaft und des Ackerbaus in der Siedlungskammer Flögeln, Niedersachsen, seit der Jungsteinzeit. Probleme der Küstenforschung im südlichen Nordseegebiet 21, Hildesheim.

Kramer, A., Bittmann, F., 2015. Revised human impact in north-western Germany during the Neolithic: methodological limits and challenges. Journal of Quaternary Science 30 (5), 434 – 451.

Mennenga, M., 2017. Zwischen Elbe und Ems. Die Siedlungen der Trichterbecherkultur in Nordwestdeutschland. Habelt, Bonn.

Plate 1. Sievern 114. 1 Charred nutshell fragment of hazel, *Corylus avellana*. 2 – 5 Charred cone scales of pine, *Pinus* sp., outer, inner, and lateral views.

2.6 Vechta 10

Site name	Vechta 10
Longitude, latitude	54.292857, 10.851122
Federal state, country	Lower Saxony, Germany
Excavation campaign	1995–2000
Excavation type	Rescue excavation
Responsible archaeologist	Jörg Eckert, Niedersächsisches Landesamt für Denkmalpflege Stützpunkt Oldenburg
Archaeological context	Grave and settlement
Analysed structures	Pits
Archaeological group	Single Grave Culture; Late Neolithic
Archaeological periodisation	Younger Neolithic (YN); Late Neolithic (LN)
Absolute dating	1900–1700 cal BCE for the Late Neolithic pit 126
Archaeobotanists	Stefanie Klooß, Elske Fischer, Kiel (SPP1400)
Archaeobotanical sampling strategy	Judgement sampling from archaeological contexts
Analysed sediment volume	375 l
Find concentrations (finds/l)	1 (0.6)
Number of carpological remains	220

Rescue excavations at Vechta 10, in Lower Saxony, were focused on a large late Bronze Age to early Iron Age urnfield cemetery of a north-west German burial tradition known as 'Schlüssellochgräber' (Eckert, 1999, 2001, 2004). The site is situated on the north-eastern periphery of this tradition. Unexpectedly, at the eastern margin of the excavation area, a Younger Neolithic burial (Single Grave Culture) and domestic structures of Late Neolithic origin were uncovered.

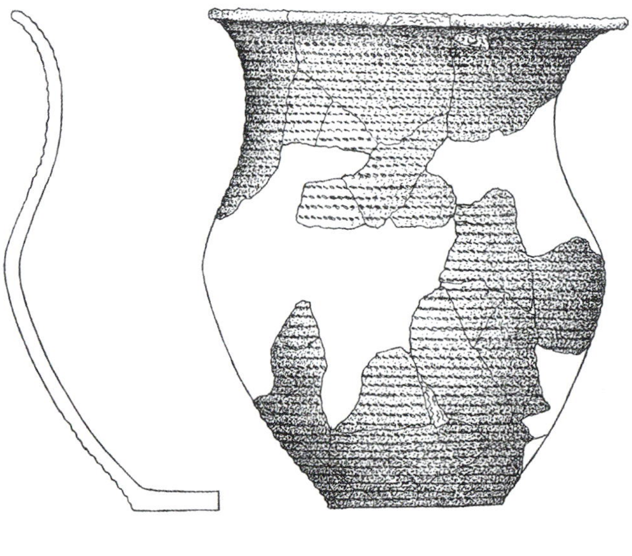

Fig. 2.6.1. All-over-ornamented Beaker, Single Grave Culture component, Vechta 10 (after Eckert, 1999: Fig. 7).

A thick-walled ceramic type known as all-over-ornamented Beaker (Fig. 2.6.1) indicates the transition from the Single Grave Culture to the Late Neolithic.

The main charred plant assemblage of about 200 cereal grains originates predominantly from one pit in a domestic context (Feature 126). A radiocarbon date (Table 8) supports the typological assignment to the Late Neolithic. Barley, mainly the free-threshing variety, is the dominant species (Table 1). It is accompanied by emmer. Typical for the Late Neolithic is the additional occurrence of gathered plants, in particular of acorn fragments. The pit in the grave from the Single Grave Culture contained only six charred remains, namely, grains of free-threshing barley and fragments of hazel nutshells. Mouse droppings were recovered from other Neolithic pits, possibly suggesting the presence of stored crops.

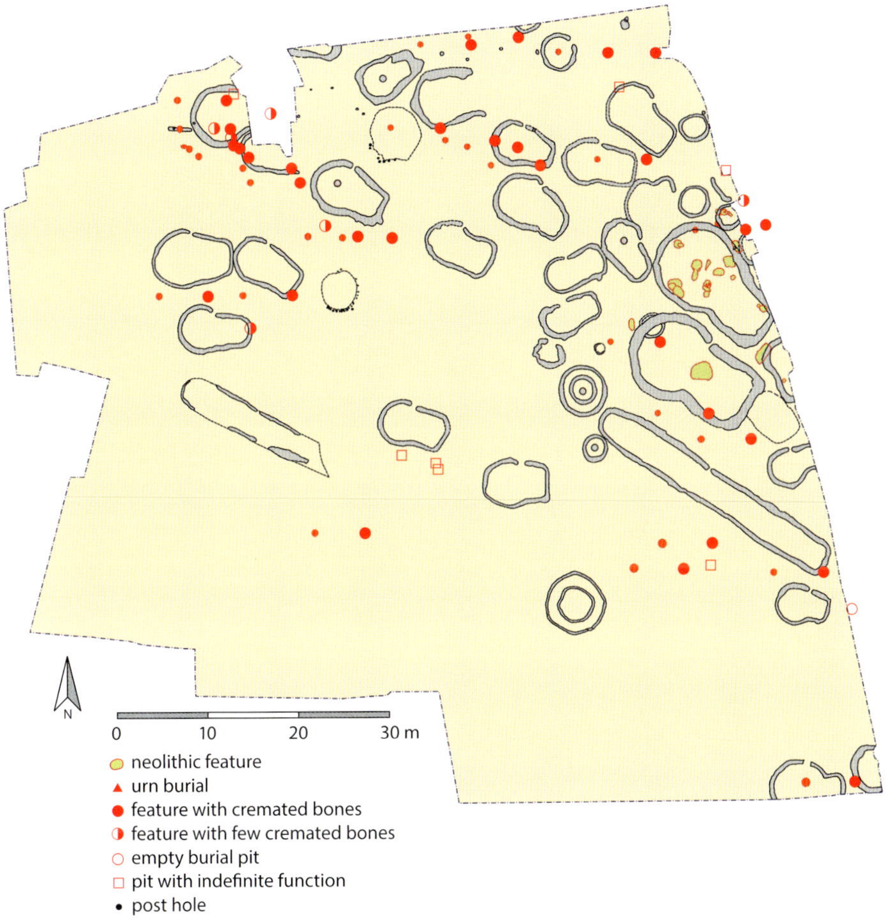

Fig. 2.6.2. Map of the excavation area, 1995 – 2000, with Neolithic features marked in green, Vechta 10 (after Eckert, 2004: Fig. 1).

Literature:

Eckert, J., 1999. Ein Urnenfriedhof der späten Bronzezeit in Vechta. Jahrbuch für das Oldenburger Münsterland 2000: 238 – 244.

Eckert, J., 2001. Von Schlüssellöchern und Langbetten. Archäologie in Niedersachsen 4, 88 – 91.

Eckert, J., 2004. Ein Gräberfeld der jüngeren Bronzezeit in Vechta, in: Fansa, M., Both, F., Haβmann, H. (Eds.), Begleitschrift zur Ausstellung Archäologie Land Niedersachsen. 25 Jahre Denkmalschutz-gesetz – 400 000 Jahre Geschichte. Archäologische Mitteilungen aus Nordwestdeutschland Beiheft 42. Theiss, Bad Langensalza, pp. 384 – 386.

Plate 1. Vechta 10. 1–3 Charred grains of free-threshing barley, *Hordeum vulgare nudum*, dorsal and ventral views. 4 – 6 Charred barley grains, *Hordeum vulgare*, dorsal and ventral views.

Plate 2. Vechta 10. 1 – 4 Charred emmer grains, *Triticum dicoccon*, dorsal and ventral/lateral views. 5 – 6 Charred acorn fragments, *Quercus* sp.

Plate 3. Vechta 10. 1 and 2 Charred nutshell fragments of hazel, *Corylus avellana*. 3 Charred mouse droppings.

3 Western Holsatia, Schleswig-Holstein: The lower moraine landscape, the outwash plain, and the western upper moraine landscape

The geomorphology of Schleswig-Holstein comprises four distinct landscapes, running north–south along the Cimbrian peninsula, all shaped during the Pleistocene. The lower moraine region of western Holsatia, the Heide-Itzehoer moraine, was formed during the Saale glaciation (c. 300 000 – 130 000 years ago). Later, the relief was transformed by periglacial processes during the Weichsel glaciation (c. 115 000 – 11 500 years ago). Soil flow and surface melting of ice leading to ablation or strong wind erosion levelled off the dynamic reliefs through transport of material from hilltops and sedimentation in depressions. Aeolian sand covered and built up other areas, while meltwaters from the Weichsel glaciers cut deeply into the Saale moraine, creating steep slopes. The lower moraine landscape in the west of Schleswig-Holstein is thus characterized by different moraine tracts interrupted by depressions that allowed for peat growth during the Holocene. It extends from the wide, peaty lowlands of the river Eider in the north, towards the river Stör in the south and the morainal apron with poor soils in the east. From c. 4500 BCE onwards into the Neolithic period, the North Sea reached the lower moraine landscape and steep cliffs, levees, and a simplification coast developed. The Wadden Sea, with its wide marshlands, was established only later, as a result of a rising sea-level with phases of transgression and regression, leading to recurrent deposition of marine sediments and the growth of peat during the late Bronze Age to Iron Age (Menke, 1988). Eastwards, a sander belt was deposited, followed

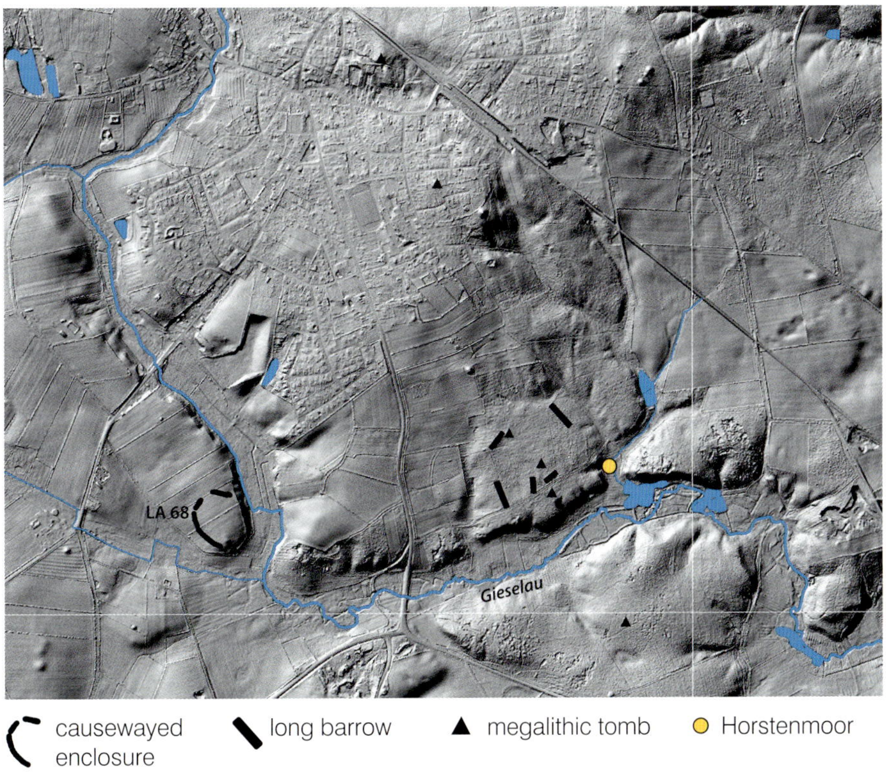

Fig. 3.1. The micro-region of Albersdorf, showing the location of Albersdorf LA68, Dieksknöll and nearby sites (after Dibbern 2016: Fig 3.1, © LVermGeo-SH).

by the hilly upper moraine landscape shaped during the Weichsel glaciation. Peat bogs serve as a natural archive for the reconstruction of the vegetation and settlement history in the lower moraine landscape (Fig. 3.1). In a peat sequence from Horstenmoor, the earliest human activity in the dense mixed-oak woodlands is observed during the late Mesolithic, with a temporary occurrence of ribwort plantain at 4500 – 4300 cal BCE. This occurrence is surprising, as it predates the onset of regular pollen finds of this anthropogenic indicator, which in other pollen sequences in the Funnel Beaker area occurs from around 3700 cal BCE. At Horstenmoor, small-scale opening up of the woodland, as shown by an increase in the light-demanding species hazel and birch, regular finds of ribwort plantain, and a decrease in elm, indicates the onset of Neolithic settlement activity. In contrast with the upper moraine area of eastern Holsatia, the elm decline in this pollen record from western Holsatia dates to 3700 cal BCE and thus is a late occurrence, one that, in addition, is clearly triggered by human activity. The intensity of human impact in the area remains stable throughout the Middle to Younger Neolithic. In the Younger to Late Neolithic, an increase in settlement activity can be observed (Feeser and Dörfler, 2016). Clusters of Neolithic sites in western Holsatia are found in the micro-regions of Albersdorf and Büdelsdorf/Borgstedt (Dibbern, 2016; Hage, 2016). They comprise burial sites, causewayed enclosures, and only few domestic sites. The archaeobotanical material thus primarily allows for insight into the ritual sphere of Neolithic societies. The micro-region of Albersdorf, in particular, shows a high density of Neolithic monuments, mainly tomb sites, such as Albersdorf LA5, Brutkamp (section 3.3), and causewayed enclosures, such as Albersdorf LA68, Dieksknöll (section 3.4). Towards the west, at what was then the edge of the North Sea, lies the domestic site of Hemmingstedt LA2 (section 3.1). The site is characterized by local flint production. New radiocarbon data on a mass find of barley from this settlement (Table 8) show that cereal was already being grown here in the Early to Middle Neolithic.

Literature:

Dibbern, H., 2016. Das trichterbecherzeitliche Westholstein. Eine Studie zur neolithischen Entwicklung von Landschaft und Gesellschaft. Habelt, Bonn.

Feeser, I., Dörfler, W., 2016. Landschaftsentwicklung und Landnutzung, in: Dibbern, H., Das trichterbecherzeitliche Westholstein. Eine Studie zur neolithischen Entwicklung von Landschaft und Gesellschaft. Habelt, Bonn, pp. 17–24.

Hage, F., 2016. Büdelsdorf/Borgstedt. Eine trichterbecherzeitliche Kleinregion. Habelt, Bonn.

Menke, B., 1988. Die holozäne Nordseeküstentransgression im Küstenbereich der südöstlichen Deutschen Bucht, in: Müller-Wille, M., Higelke, B., Hoffmann, D., Menke, B., Brande, A., Bokelmann, K., Saggau, H.E., Kühn, H.J. (Eds.), Norderhever-Projekt. 1 Landschaftsentwicklung und Siedlungsgeschichte im Einzugsgebiet der Norderhever (Nordfriesland). Offa-Bücher 66. Wachholtz, Neumünster.

3.1 Hemmingstedt LA2

Site name	Hemmingstedt LA2
Longitude, latitude	54.170047, 9.085104
Federal state, country	Schleswig-Holstein, Germany
Excavation campaign	1984–1986
Excavation type	Research excavation (prior to construction of A23 motorway)
Responsible archaeologist	Ingo Clausen, Archäologisches Landesamt Schleswig-Holstein
Archaeological contexts	Settlement, two-aisled houses, ard marks
Analysed structures	Pits, postholes
Archaeological group	Funnel Beaker Culture
Archaeological periodisation	Early Neolithic (EN I b–II)
Absolute dating	3703–3319 cal BCE (according to new dates from the SPP1400)
Archaeobotanists	Stefanie Klooß, Wiebke Kirleis, Kiel (SPP1400)
Archaeobotanical sampling strategy	Judgement sampling from archaeological contexts
Analysed sediment volume	est. 280 l (extrapolated, flotation with 1 mm sieves on-site)
Find concentrations (finds/l)	1
Number of carpological remains	370

The settlement site of Hemmingstedt LA2, occupied from EN II to MN I, was a coastal site at the edge of the North Sea. In this period, the sea reached the upper moraine, and the sea marshes did not yet exist. Towards the sea, a simplification coast existed, with cliffs, debris strand, levees, and spits that cut off shallow coastal lakes in the interior. Hemmingstedt LA2 was situated on a peninsula at the eastern margin of a former coastal lake basin that developed into the peat bog Liether Moor. The site is characterized by activity areas for flint production (Clausen, 1994). The flint was a local resource that was gathered at the shoreline of the North Sea, at a distance of 1.6 km from the settlement (Krause-Kyora, 2008). The distribution of postholes allows for the reconstruction of a two-aisled house of the Dagstorp II type (Fig. 3.3.1). Criss-cross ard marks were discovered over an area of about 1700 m² overtop of the domestic features dating to EN II to MN I; these marks thus date later. Nine conventional radiocarbon dates on charcoal suggest a late Middle Neolithic date for the site (Krause-Kyora, 2007: 84).

New AMS dating in 2014 of barley grains from a pit with domestic remains (Feature 298) supports the dating to EN II to MN I, and that of a nutshell fragment from a posthole with only two plant remains (Feature 208) indicates EN I (Table 8). The charred plant material thus has to be considered to mainly relate to EN II to MN I (Table 2). Free-threshing barley dominates the plant assemblage. Only five single grains can be attributed to emmer. Ten nutshell fragments from hazel represent the gathered plants. Weed species are rare and are not documented here. They comprise two single finds, one of black bindweed and one of vetch. In addition to some culm fragments, eight cylindric charred remains in the size class of cereal grains were found, but these were identified as mouse droppings.

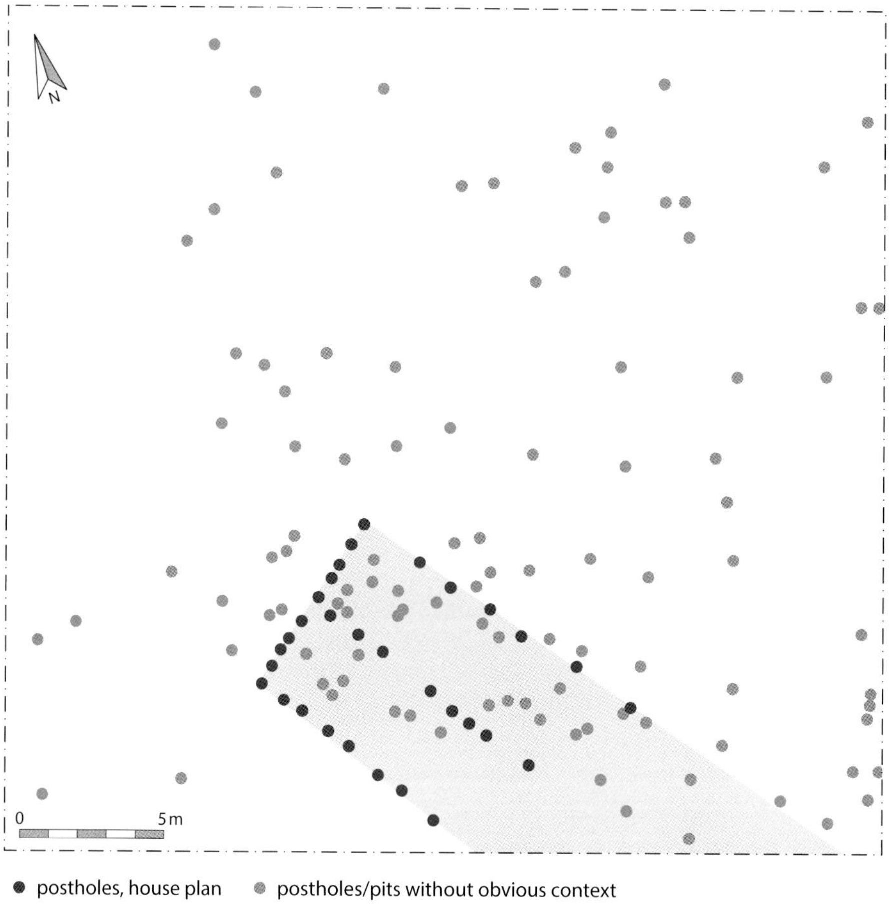

Fig. 3.1.1. Distribution of postholes and reconstruction of a Dagstorp II type two-aisled house, Hemmingstedt LA2 (after Dibbern, 2016: Fig. 9.3).

Literature:

Clausen, I., 1994. Hemmingstedt, Kr. Dithmarschen. Neolithische Siedlung LA2. Offa 51, 221–228.

Dibbern, H., 2106. Das trichterbecherzeitliche Westholstein. Eine Studie zur neolithischen Entwicklung von Landschaft und Gesellschaft. Habelt, Bonn.

Krause-Kyora, B., 2007. Ein Trümmerhaufen: Lithische Aspekte aus der neolithischen Siedlung Hemmingstedt LA2. Unpublished Master's thesis, Kiel University.

Krause-Kyora, B., 2008. Ein Trümmerhaufen, Steinzeitliches Feuersteinhandwerk und der Feuersteinhandel an der Nordseeküste. Aspekte aus der neolithischen Siedlung Hemmingstedt, Kr. Dithmarschen. Archäologie in Schleswig/Arkæologi i Slesvig 12, 39–52.

Müller, J., 2017. Großsteingräber, Grabenwerke, Langhügel. Frühe Monumentalbauten Mitteleuropas. Theiss, Darmstadt.

Plate 1. Hemmingstedt LA2. 1 Charred grains of free-threshing barley, *Hordeum vulgare nudum*. 2 Charred barley grains, *Hordeum vulgare*.

Plate 2. Hemmingstedt LA2. 1 Charred emmer grain, *Triticum dicoccon*, dorsal, ventral, and lateral views. 2 and 3 Charred nutshell fragments of hazel, *Corylus avellana*.

3.2 Borgstedt LA35

Site name	Borgstedt LA35
Longitude, latitude	54.331484, 9.709208
Federal state, country	Schleswig-Holstein, Germany
Excavation campaign	2009
Excavation type	Rescue excavation (prior to construction of A23 motorway)
Responsible archaeologist	Cornelia Lübke, Archäologisches Landesamt Schleswig-Holstein
Archaeological context	Burial mound with central megalithic burial chamber
Analysed structures	Burial mound and burial chamber
Archaeological group	Funnel Beaker Culture, Single Grave Culture, Bell Beaker phenomenon
Archaeological periodisation	Middle Neolithic to Late Neolithic (MN–LN)
Absolute dating	c. 3300–1700 cal BCE
Archaeobotanists	Stefanie Klooß, Wiebke Kirleis, Kiel (SPP1400)
Archaeobotanical sampling strategy	Judgement sampling from archaeological contexts
Analysed sediment volume	c. 139 l
Find concentrations (finds/l)	1 (0.8)
Number of carpological remains	115

Fig. 3.2.1. Bird's eye view of the excavation, facing south, Borgstedt LA35 (photo: ALSH, Schleswig).

Fig. 3.2.2. Reconstruction of the Middle to Late Neolithic landscape, showing the river Eider and the enclosure and burial ground in the micro-region of Büdelsdorf/ Borgstedt (drawing: Susanne Beyer, UFG Kiel).

Borgstedt is situated in the transition zone between the hilly upper moraine landscape and the sandier area that was shaped during the Weichsel glaciation at the former edge of the ice shield from the Fockbeker advance. The site Borgstedt LA35 was discovered in the rural district of Rendsburg-Eckernförde prior to construction work on a house (Fig. 3.2.1). It belongs to a Neolithic micro-region with a burial ground with numerous monumental tombs oriented towards a Funnel Beaker enclosure site 800 m to the south-west that is located on a spur into the river Eider (Fig. 3.2.2; Hage, 2016). Borgstedt LA35 consists of a grave mound with stone circles with an overall diameter of 24.5 m (Lübke, 2010). The mound was erected in three phases, dating to the Middle Neolithic Funnel Beaker Culture, the Younger Neolithic Single Grave Culture, and the Bell Beaker phenomenon, at the transition from the Late Neolithic to the Early Bronze Age. The first phase of the grave mound consists of a megalithic grave chamber that was covered with soil in the Middle Neolithic. Numerous silex artefacts, including fragments of sickles, were found on the mound's surface. The finds show that domestic activities took place at the grave mound in the Younger to Late Neolithic.

The archaeobotanical assemblage also shows a composition that is typical for domestic sites (Table 2). Grains of barley, emmer, and einkorn, as well as emmer glume bases, hint at cereal processing. With knotweed, white goosefoot, common sheep sorrel, and ribwort plantain, a typical Middle to Late Neolithic weed flora is present, again showing that impurities were being removed from cereals before cooking. Sheep sorrel is indicative for poor soils in the site's vicinity, although a single occurrence does not constitute robust evidence. The assemblage further contains nutshell fragments from hazel, showing that the occupants gathered plants in order to gain nutritious supplemental foods throughout the Neolithic.

Literature:

Hage, F., 2016. Büdelsdorf/Borgstedt. Eine trichterbecherzeitliche Kleinregion. Habelt, Bonn.

Lübke, C. 2010. Von Häusern und Gräbern. Bauliche Konstruktionen. Archäologische Nachrichten aus Schleswig-Holstein 2010, 41–45.

Plate 1. Borgstedt LA35. 1 Charred emmer grains, *Triticum dicoccon*, dorsal view. 2 – 3 Charred emmer grains, lateral and ventral views. 4 Charred glume bases of emmer. 5 Charred grains of free-threshing barley, *Hordeum vulgare nudum*, dorsal, lateral, and ventral views.

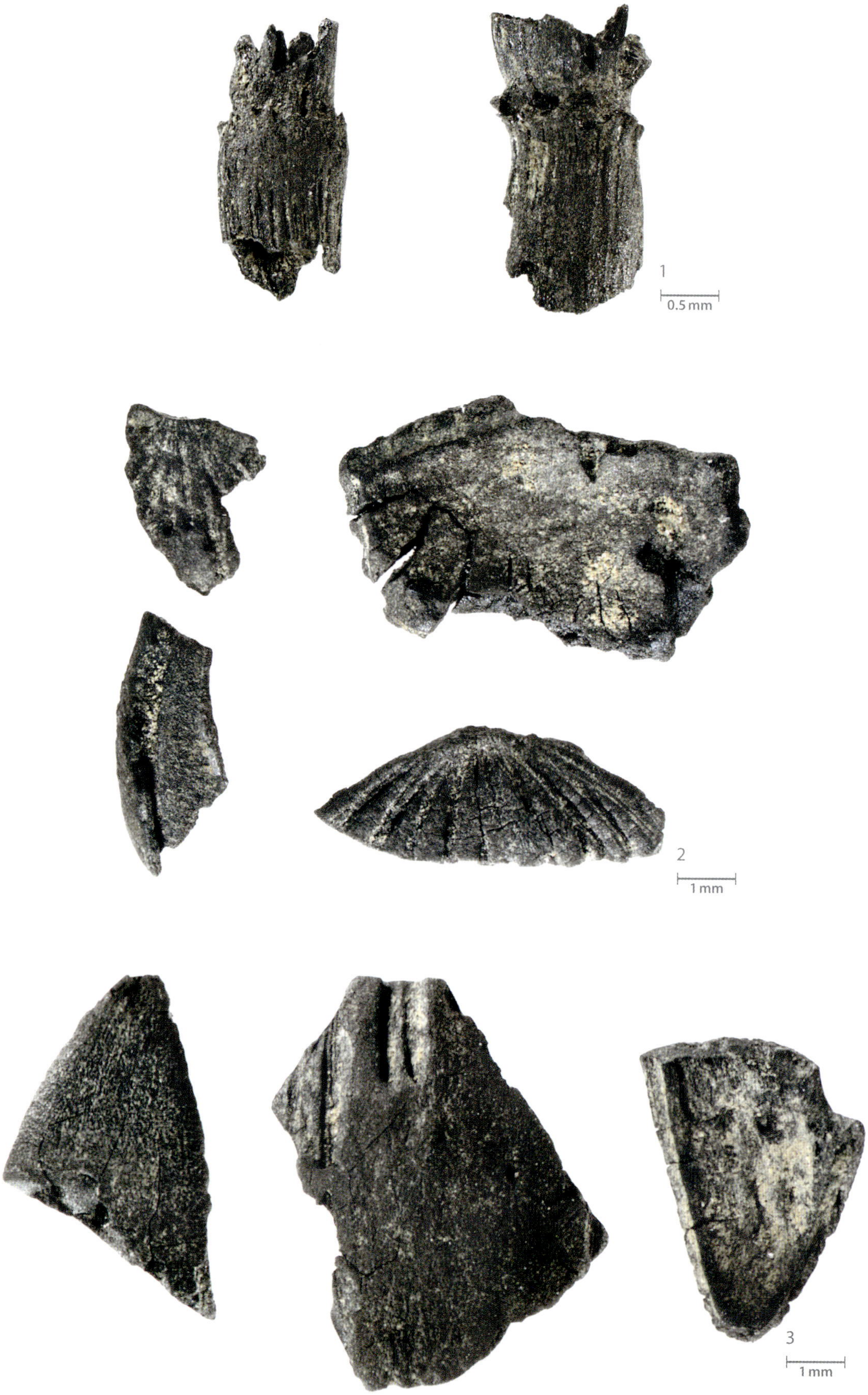

Plate 2. Borgstedt LA35. 1 Charred node of the grass family (cereal?), Poaceae. 2 and 3 Charred nutshell fragments of hazel, *Corylus avellana*.

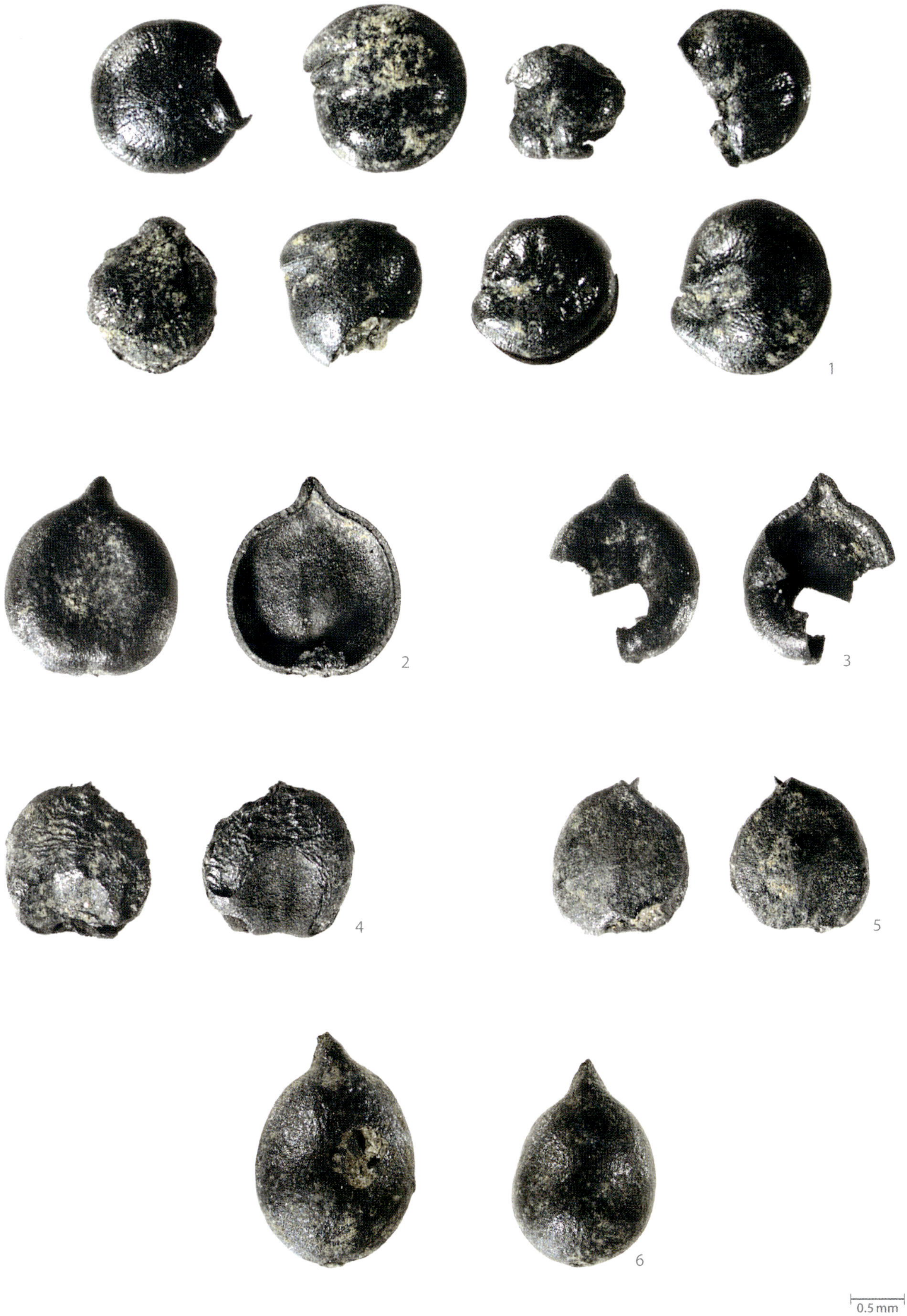

Plate 3. Borgstedt LA35. 1 Charred seeds of white goosefoot, *Chenopodium album*. 2 and 3 Charred curlytop knotweed, *Persicaria lapathifolia*, nutshells, outer and inner views. 4 and 5 Charred curlytop knotweed, inner fruits, front and back views. 6 Charred fruits of lady's thumb, *Persicaria maculosa*.

Plate 4. Borgstedt LA35. 1 Charred fruit of common sheep sorrel, *Rumex acetosella*. 2 – 4 Charred seeds of ribwort plantain, *Plantago lanceolata*, ventral and dorsal views.

3.3 Albersdorf LA5, Brutkamp

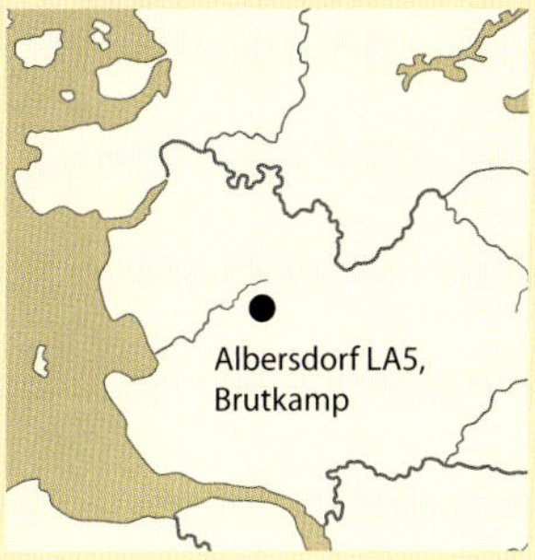

Site name	Albersdorf LA5, Brutkamp
Longitude, latitude	54.143663, 9.290806
Federal state, country	Schleswig-Holstein, Germany
Excavation campaign	2009
Excavation type	Research excavation (SPP1400)
Responsible archaeologists	Hauke Dibbern, Franziska Hage, Johannes Müller, Kiel
Archaeological context	Megalithic tomb (dolmen)
Analysed structures	Entrance area and soil mantle of the grave mound
Archaeological groups	Funnel Beaker Culture; Dagger groups
Archaeological periodisation	Early Neolithic and Late Neolithic (EN II and LN I)
Absolute dating	3600–3530 and 2280–2030 cal BCE
Archaeobotanists	Henrike Röhrs, Stefanie Klooß, Wiebke Kirleis, Kiel (SPP1400)
Archaeobotanical sampling strategy	Systematic grid-wise sampling, one sample per quadrant in each cut layer of 20 cm, plus subjective sampling of features
Analysed sediment volume	c. 470 l
Find concentrations (finds/l)	1 (0.4)
Number of carpological remains	90

The megalithic tomb site of Albersdorf LA5, Brutkamp, is situated in the micro-region of Albersdorf, in Dithmarschen, Schleswig-Holstein. The monument is an extended polygonal dolmen of five orthostats located in the centre of a round burial mound that is surrounded by a stone circle. The burial chamber is covered by a boulder weighing about 18 t (Fig. 3.3.1). In summer 2009, the entrance area of the grave, on the narrow south-east side, was excavated within an east–west oriented grid of 12 quadrants in the frame of the priority program SPP1400 'Early Monumentality and Social Differentiation' (Fig. 3.3.2) (Dibbern, 2016; Roehrs et al., 2013). The construction of the dolmen took place in the Early Neolithic, 3600–3530 cal BCE (Dibbern, 2016: 83). It was in use mainly during the Middle Neolithic. A second soil mantle over the grave mound and the stone package date to the Late Neolithic Dagger period (2280–2030 cal BCE). During the re-opening episode in Late Neolithic times, the contents of the burial chamber were presumably cleared into the entrance area (Dibbern, 2016: 98; Müller, 2017: 78). Inhumation is the common burial practice for Funnel Beaker and Late Neolithic groups, but no bones were preserved at the site due to the dry soil conditions. In addition, the grave chamber was probably affected by modern disturbance, but secondary burials were not identified.

Only charred plant material has been preserved (Table 2), and this shows a predominance of collected plants, in particular nutshells of hazel (Kirleis

Fig. 3.3.1. The Megalithic tomb at Albersdorf LA5, Brutkamp (photo: Sara Jagiolla, UFG Kiel).

and Klooß, 2014). Based on the radiocarbon dates, the botanical remains could be separated into two phases (Table 8). An Early to Middle Neolithic group with 77 finds includes hazel nutshells, raspberry seeds, immature lime fruits, knotweed, cleavers, and common nipplewort. A Late Neolithic group with 43 remains comprises nutshells of hazel, bulbs of tuber oat grass *Arrhenatherum elatius* var. *bulbosum*, raspberry seeds, and blackthorn. The percentage of bulbs of tuber oat grass as a proportion of all plant remains from the Late Neolithic is unusually high, reaching 30 per cent. Two radiocarbon dates (Table 8) indicate that the bulbs were brought into the

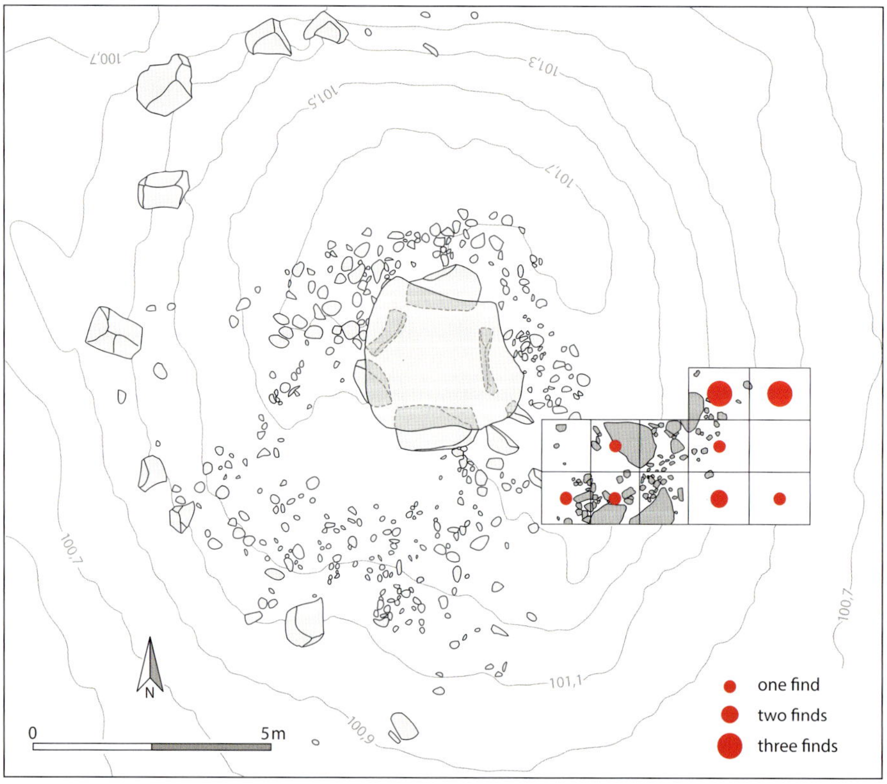

Fig. 3.3.2. The Megalithic tomb at Albersdorf LA5, Brutkamp: excavated area and distribution of charred bulbs of tuber oat grass (after Roehrs et al., 2013: Fig. 2).

Fig. 3.3.3. Tuber oat grass, *Arrhenatherum elatius* var. *bulbosum* in April 2016. Bulbs, partly chained, above ground in/above the litter zone (photo: Wiebke Kirleis, UFG Kiel).

grave context concurrent with dagger fragments during the Late Neolithic (2250 – 2000 cal BCE) and are therefore not contemporaneous with the construction phase of the megalithic tomb. The scattered occurrence of the bulbs within the upper soil mantle implies they have no direct connection to the cleared-out grave inventory, so their presence as gift or ritual symbol can be excluded. The occurrence of the sweet grass *A. elatius* var. *bulbosum* surely shows that the surroundings of the grave during the Late Neolithic were an open landscape, where derelict land developed that was used neither for pasture nor for agriculture. *Arrhenatherum* may have been brought into the grave complex as bulbs together with the soil material for the mantle after burning of the grass at the place of soil extraction, or *Arrhenatherum* may have been growing on the grave mound itself during the Late Neolithic and was burned before a possible re-opening of the entrance in connection with a secondary burial or other ritual activity (Roehrs et al., 2013).

In recent times, the bulbous subspecies of *Arrhenatherum elatius* has become extinct in Schleswig-Holstein. In prehistory, a decline in tuber oat grass is observed in northern central Europe starting in the pre-Roman Iron Age (Roehrs et al., 2013). This decline may have been induced by a technological innovation, namely, the introduction of the soil-turning mouldboard plough, because the growth of *Arrhenatherum elatius* var. *bulbosum* is heavily affected by soil disturbance, as has recently been shown by growing experiments at Kiel botanical garden with breeds from France and Spain from the genebank of the Leibniz-Institut für Pflanzengenetik und Kulturpflanzenforschung (IPK) Gatersleben (GR 5517, GR 7127, and GR 7133).

Literature:

Dibbern, H., 2106. Das trichterbecherzeitliche Westholstein. Eine Studie zur neolithischen Entwicklung von Landschaft und Gesellschaft. Habelt, Bonn.

Kirleis, W., Klooß, S., 2014. More than simply fallback food? Social context of plant use in the northern German Neolithic, in: Chevalier, A., Marinova, E., Peña-Chocarro, L. (Eds.), Plants and people: choices and diversity through time. Earth Series 1. Oxbow Books, Oxford, pp. 326 – 335.

Müller, J., 2017. Großsteingräber, Grabenwerke, Langhügel. Frühe Monumentalbauten Mitteleuropas. Theiss, Darmstadt.

Roehrs, H., Klooβ, S., Kirleis, W., 2013. Evaluating prehistoric finds of *Arrhenatherum elatius* var. *bulbosum* in north-western and central Europe with emphasis on the first Neolithic finds in Northern Germany. Journal of Archaeological and Anthropological Sciences 5(1), 1–15.

Plate 1. Albersdorf LA5, Brutkamp. 1 Charred grain of barley, *Hordeum vulgare*, dorsal, lateral, and ventral views. 2 – 6 Charred nutshell fragments of hazel, *Corylus avellana*. 7 Charred hazel nut, unripe. 8 and 9 Charred fragment of a stone kernel of blackthorn, *Prunus spinosa*, outer and inner views.

Plate 2. Albersdorf LA5, Brutkamp. 1 and 2 Charred seed of raspberry, *Rubus idaeus*. 3–8 Charred bulbs of tuber oat grass, *Arrhenatherum elatius* var. *bulbosum*. 9 Charred fruit of lime, *Tilia* sp.

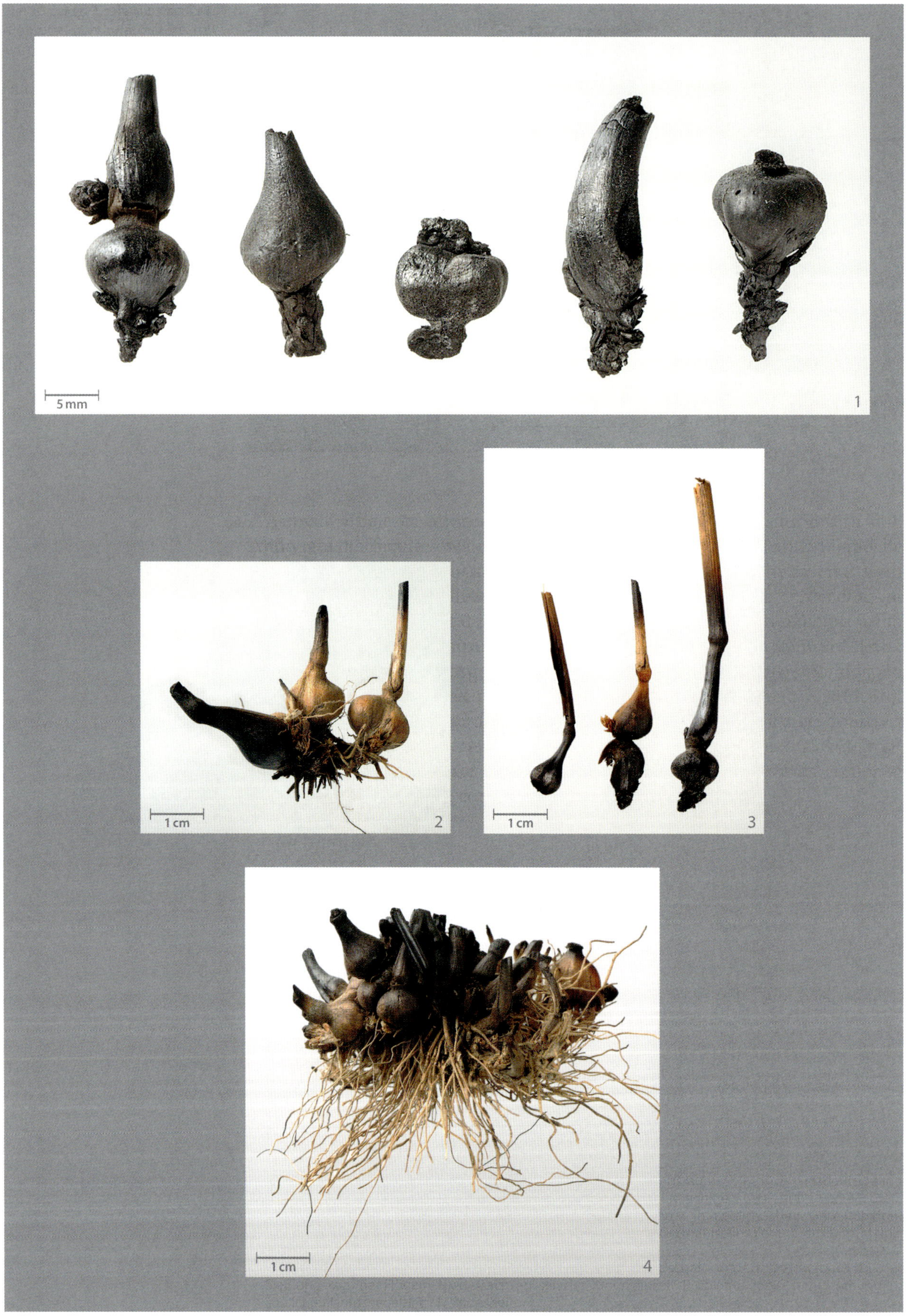

Plate 3. 1 Charred modern bulbs of tuber oat grass, *Arrhenatherum elatius* var. *bulbosum*. 2 – 4 Partly charred lumps of modern bulbs of tuber oat grass (photos: Sara Jagiolla, UFG Kiel).

3.4 Albersdorf LA68, Dieksknöll

Site name	Albersdorf LA68, Dieksknöll
Longitude, latitude	54.133198, 9.278570
Federal state, country	Schleswig-Holstein, Germany
Excavation campaign	2010
Excavation type	Research excavation (SPP1400)
Responsible archaeologists	Hauke Dibbern, Johannes Müller, Kiel
Archaeological context	Causewayed enclosure
Analysed structures	Ditch, pits
Archaeological groups	Funnel Beaker Culture, Single Grave Culture
Archaeological periodisation	Early Neolithic to Younger Neolithic (EN Ib–YN 2)
Absolute dating	3760–2500 cal BCE
Archaeobotanists	Stefanie Klooß, Wiebke Kirleis, Kiel (SPP1400)
Archaeobotanical sampling strategy	Systematic sampling of all features and layers
Analysed sediment volume	c. 1480 l
Find concentrations (finds/l)	1 (0.1)
Number of carpological remains	91

The site of Albersdorf LA68, Dieksknöll, is situated in the micro-region of Albersdorf, a lower moraine landscape in the district of Dithmarschen, western Holsatia. In the Neolithic, this micro-region was located in the hinterland of the coastal area of the North Sea, since the marshlands did not yet exist and the North Sea directly reached the steep cliffs of the lower moraine, where levees and a simplification coast were formed. The landscape in the micro-region of Albersdorf is characterized by the river valley of the Gieselau, which is cut deeply into the lower moraine landscape. The river Gieselau marks the south-western border of the micro-region, which shows a slightly rising terrain towards the north-east, reaching an absolute height of 60 m above sea-level. The micro-region is located in a geographically favourable landscape. In prehistoric times, the river Gieselau was navigable. Since it was connected with the river Eider, which discharges into the North Sea, this waterway provided possible access to the sea. The area shows a high density of Neolithic sites (Fig. 3.1). The causewayed enclosure site Albersdorf LA68 was established in the thirty-eighth century cal BCE on the spur called Dieksknöll, facing towards the rivers Gieselau and Forbek. It consists of a segmented ditch surrounding an empty space of 2.5 ha and a wooden palisade. Different kinds of activities have been reconstructed for the ditch areas. The ditch system was recurrently revisited throughout the time period 3760–2500 cal BCE. Initially, it was dug out and refilled every second to third generation and from 3400 cal BCE onwards only every sixth to eight generation (Dibbern, 2016: 35;

Fig. 3.4.1. View of the excavation, Albersdorf LA68, Dieksknöll (photo: Wiebke Kirleis, UFG Kiel, March 2010).

Müller, 2017: 84). Pottery and stone artefacts were intentionally deposited. One ceramic plate with impressed edges from the ditch shows several imprints of plant parts. A radiocarbon date of 3650 cal BCE from the layer covering the ceramic plate serves as a *terminus ante quem*. Imprints of organic material are typical features of this kind of ceramic plate with impressed edges, which are mainly distributed in the western parts of central Europe during the Neolithic. Their occurrence goes along with the phenomenon of the causewayed enclosures, which links the northern European Funnel Beaker Culture with the Michelsberg/Chasséen Culture farther to the south-west, where causewayed enclosures as expressions of monumentality were constructed starting in the late fifth millennium BCE (Fig. 3.4.2). The plant imprints on the plate with the impressed edges from the site Albersdorf LA68, Dieksknöll, constitute the first evidence for free-threshing wheat with tetraploid genomic constitution, *Triticum durum/turgidum*, for the northern European Funnel Beaker Culture. It was known from the occurrence of single grains that one of the free-threshing wheats supplemented the cereal spectrum to a minor content. However, the two species, tetraploid and hexaploid, can only be separated morphologically if rachis segments are preserved. The distinction of tetraploid rachis segments is possible because they show prominent swellings below the glume insertion points, a straight-sided trapezoid shape with a maximum width in the upper part, and an absence of longitudinal lines near the outer edge of the rachis internode in abaxial view (Hillman et al., 1996; Jacomet, 2006).

The archaeobotanical investigations on the charred plant remains from the soil samples from the ditch reveal evidence for plants from the neighbouring carr forest, in particular, vegetative remains and cone scales from alder (Table 2). There is no evidence for crop species and only minor evidence for gathered plants, such as raspberry and bramble (Klooβ and Kirleis, 2012). Thus, the archaeobotanical evidence does not hint at daily, routine activities, such as crop processing and cooking. Rather, it supports the archaeological assumption, based on a lack of evidence for secular use in the material culture, that this causewayed enclosure functioned as a ritual site.

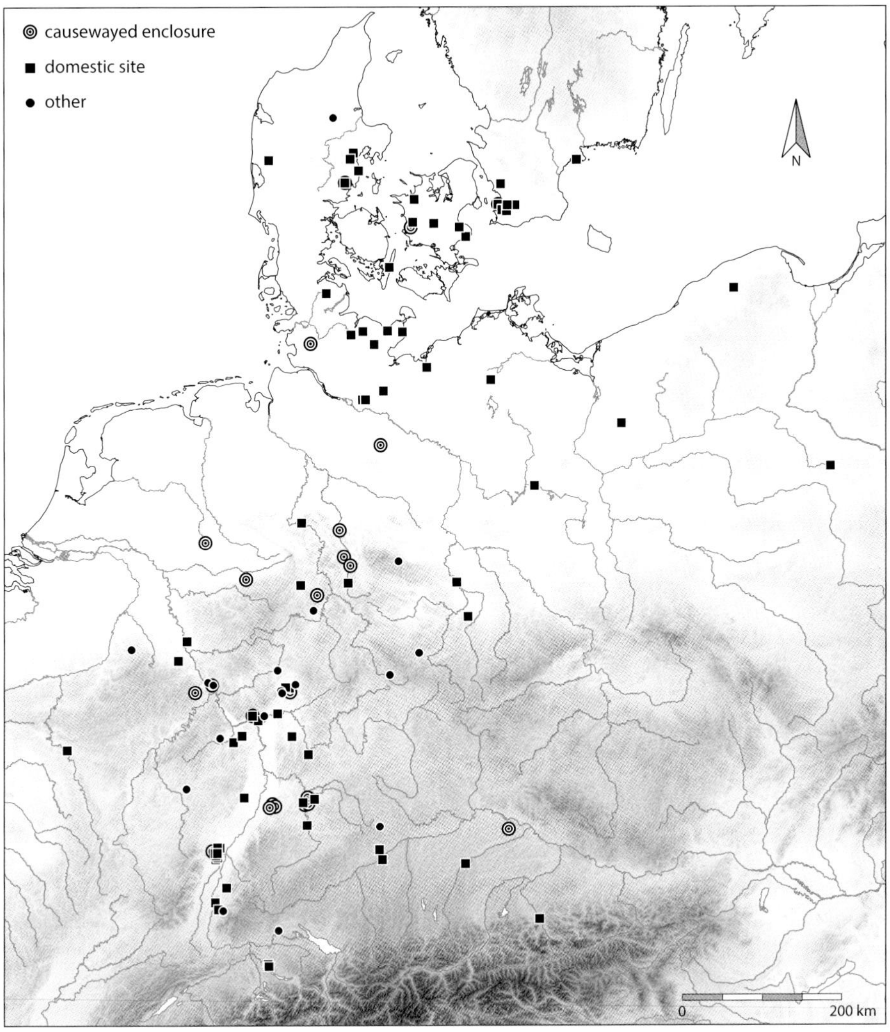

Fig. 3.4.2. Distribution of ceramic plates with impressed edges in central Europe (after Dibbern, 2016: Fig. 6.32).

Literature:

Dibbern, H., 2016. Das trichterbecherzeitliche Westholstein. Eine Studie zur neolithischen Entwicklung von Landschaft und Gesellschaft. Habelt, Bonn.

Hillman, G.C., Mason, S., De Moulins, D., Nesbitt, M., 1996. Identification of archaeological remains of wheat: the 1992 London workshop. Circaea 12, 195–210.

Jacomet, S., 2006. Identification of cereal remains from archaeological sites, fourth ed. Archaeobotany Lab IPAS, Basel University, Basel. https://duw.unibas.ch/fileadmin/user_upload/duw/IPNA/PDF_s/PDF_s_in_use/Cereal_Id_Manual_engl.pdf

Kirleis, W., Fischer, E., 2014. Neolithic cultivation of tetraploid free threshing wheat in Denmark and northern Germany: implications for crop diversity and societal dynamics of the Funnel Beaker Culture. Vegetation History and Archaeobotany 23, Supplement 1, 81–96.

Klooß, S., Kirleis, W., 2012. Die verkohlten Pflanzenreste aus dem trichterbecherzeitlichen Erdwerk Albersdorf-Dieksknöll, Kr. Dithmarschen, Schleswig-Holstein, in: Hinz, M., Müller, J., Lüth, F. (Eds.), Frühe Monumentalität und soziale Differenzierung 2. Habelt, Bonn, pp. 297–302.

Plate 1. Albersdorf LA68, Dieksknöll. 1 Ceramic plate with organic imprints, GE4. 2 Detail of 1. 3 Imprints of rachis segments of tetraploid free-threshing wheat, *Triticum durum/turgidum,* in the plate with the impressed edges. 4 Plastic cast made with the dental casting compound Vinyl Polysiloxane (Panasil contact plus, x-light) of the imprints of rachis segments of tetraploid free-threshing wheat. 5 Imprints of a spikelet of barley, *Hordeum vulgare*. 6 Imprints of straw fragments.

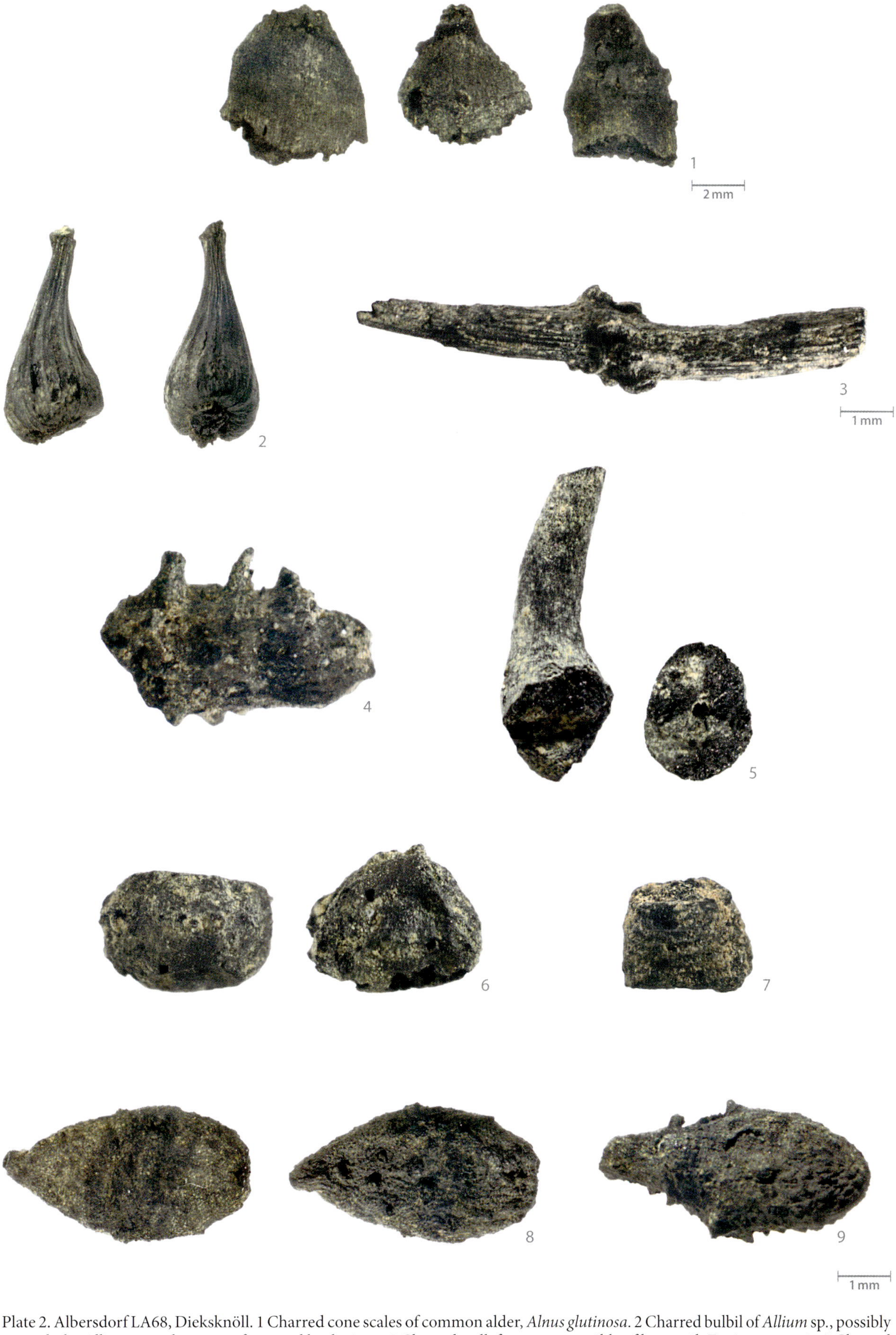

Plate 2. Albersdorf LA68, Dieksknöll. 1 Charred cone scales of common alder, *Alnus glutinosa*. 2 Charred bulbil of *Allium* sp., possibly rocambole, *Allium scorodoprasum*, front and back views. 3 Charred stalk fragment, possibly of horsetail, *Equisetum* sp. 4 – 9 Charred root tubers (photos: Stefanie Klooß, UFG Kiel).

4 South-eastern Holsatia, Schleswig-Holstein: The upper moraine hilly landscape

The upper moraine hilly landscape of south-eastern Holsatia, in Schleswig-Holstein, is situated in the eastern hilly country that was formed during the Weichsel glaciation, in the period 115 000 – 11 500 BCE. The upper moraine landscape in the coastal hinterland shows a dynamic geomorphology with eroded summits, slopes, and depressions, as well as areas with only slight relief.

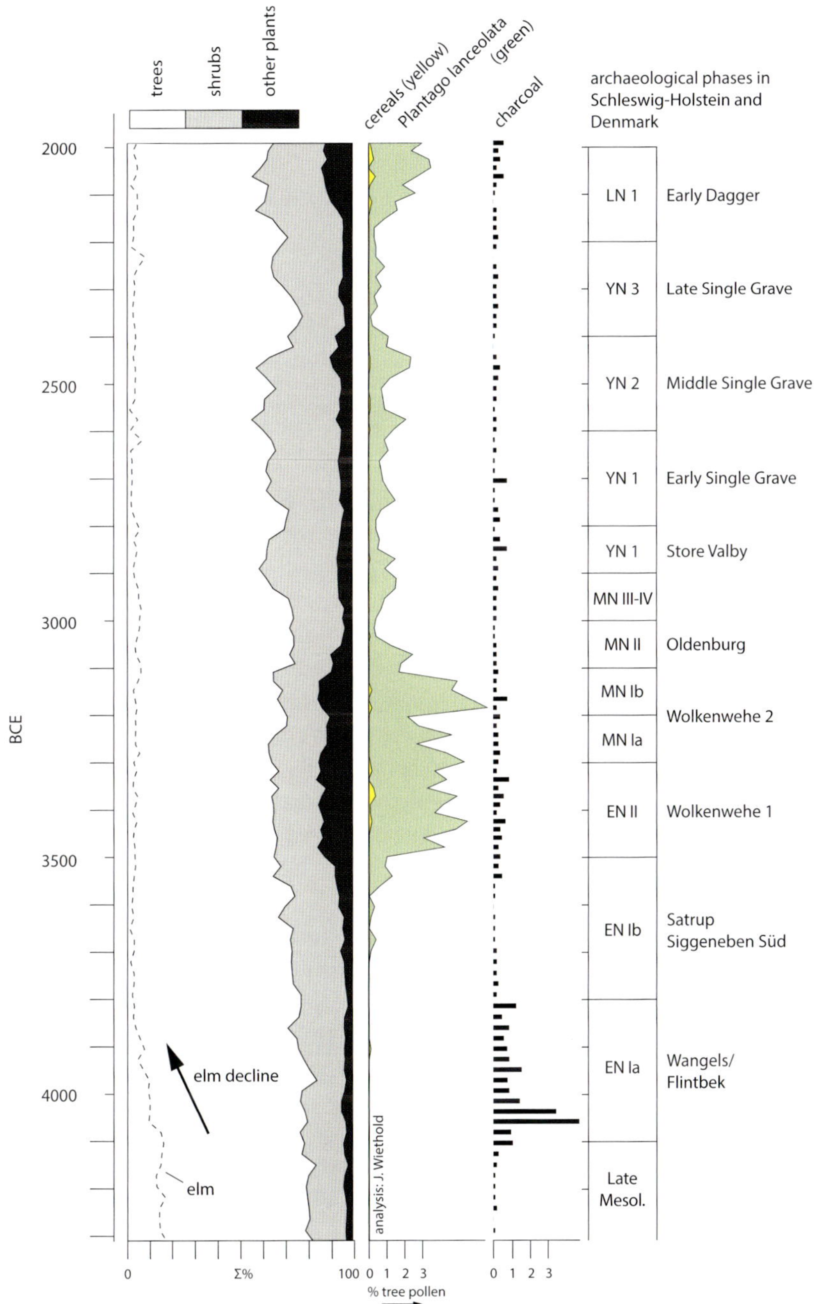

Fig. 4.1. Simplified pollen diagram from Lake Belau (after Kirleis et al., 2011: 33).

Fig. 4.2. Land that has been left fallow for one year, dominated by ribwort plantain, close to Kosel, Schleswig-Holstein (photo: Agnes Heitmann, UFG Kiel, 9 July 2015).

The landscape of the south-eastern part of Holsatia, south of the Kiel fjord, towards the town of Bad Oldesloe in the south, is characterized by numerous bigger lakes, such as the Selenter See, the Lanker See, and the lakes in the Plön area. The main soil substratum is boulder clay, and the top soil is luvisol. In the Neolithic, the area was densely forested. Pollen analytical and sedimentological investigations with high temporal resolution on a sediment sequence from Lake Belau describe the vegetation and settlement history of the region (Dörfler et al., 2012). The pollen assemblages for the Neolithic period show a distinct local signal that allows reconstruction of a stepwise establishment of farming (Fig. 4.1). Around 4050 BCE, over a period of around 50 years, recurrent late frost events are indicated in the sediment record (Dreibrodt et al., 2012). It is only after this period that the dense woodlands were opened up by use of fire and the elm declined in frequency relative to the other deciduous trees. In this first phase, small-scale crop growing took place on garden-like plots in the immediate vicinity of the farmsteads. In the late Early Neolithic, from 3600 BCE onwards, a shift in the subsistence regime can be observed. An expansion of woodland openings accompanies crop growing on large arable fields, as shown by an increase in perennial weed species. In particular the ribwort plantain hints to less intense weeding and as well to possible interposed fallow periods on the fields (illustrated by Figs. 4.2 and 4.3). The establishment of large arable fields was triggered by the technological innovation of the ard, which had been powered by draught cattle on a regular basis since the Middle Neolithic, from 3300 BCE onwards (Johannsen, 2006). The use of a cumbersome, animal-powered ard accounts for fields that have been cleared from big boulders and stumps. While this precondition for farming had already been met with respect to the boulders because boulders were being removed to erect megalithic tombs since around 3650 BCE, the uprooting and clearing of the fields was a challenge for the farmers. Neolithic sites in south-eastern Holsatia comprise burial sites, causewayed enclosures, and domestic sites, with the latter being extremely rare. The most prominent burial site of the area is the megalithic cemetery of Flintbek (section 4.8). Intensive radiocarbon dating established a high temporal resolution biography of the monuments (Mischka, 2011). The oldest absolute dates for cart tracks in the north derive from this site, and ard marks from the site indicate the use of

Fig. 4.3. Detail of ribwort plantain in the fallow plot shown in Fig. 4.2 (photo: Agnes Heitmann, UFG Kiel, 9 July 2015).

this arable tool in the late Early Neolithic. The causewayed enclosures in this area yielded diverse plant assemblages. Rastorf LA73 (section 4.2) comprises remains of wild plants, similar to the site of Albersdorf LA68, Dieksknöll (section 3.4), whereas cereal remains at the enclosure site of Bad Segeberg LA94 (section 4.6) may indicate that daily routine activities were carried out. The function of Neolithic causewayed enclosures is a widely debated topic (e. g. Klassen, 2014; Meyer and Raetzel-Fabian, 2006). From an archaeobotanical point of view, there does not appear to be one main explanation for this type of archaeological context. It seems more likely that each causewayed enclosure has its own biography. Important insight into the plant economy from domestic contexts in south-eastern Holsatia comes from Rastorf LA6a/c (section 4.1) for the Early Neolithic and Bosau LA242 (section 4.7) for the Late Neolithic. While Rastorf LA6a/c provides us with rare and early evidence for the use of linseed (*Linum usitatissimum*), Bosau LA242 is exceptional because it is one of the very few domestic sites from the Late Neolithic in northern Germany.

Literature:

Dörfler, W., Feeser, I., Bogaard, C., Dreibrodt, S., Erlenkeuser, H., Kleinmann, A., Merkt, J., Wiethold, J., 2012. A high-quality annually laminated sequence from Lake Belau, northern Germany: revised chronology and its implications for palynological and tephrochronological studies. The Holocene 22, 1413–1426. https://doi.org/10.1177/0959683612449756

Dreibrodt, S., Zahrer, J., Bork, H.-R., Brauer, A., 2012. Witterungs- und Umweltgeschichte während der norddeutschen Trichterbecherkultur, rekonstruiert auf Basis mikrofazieller Untersuchungen an jahresgeschichteten Seesedimenten, in: Hinz, M., Müller, J. (Eds.), Siedlung, Grabenwerk, Großsteingrab. Studien zu Gesellschaft, Wirtschaft und Umwelt der Trichterbechergruppen im nördlichen Mitteleuropa. Frühe Monumentalität und Soziale Differenzierung. 2. Habelt, Bonn, 145–158.

Johannsen, N. N., 2013. Draught cattle and the South Scandinavian economies of the 4th millennium BC. Environmental Archaeology, 11, 35–48.

Kirleis, W., Feeser, I., Klooß, S., 2011. Umwelt und Ökonomie. Archäologie in Deutschland 2, 32–35.

Klassen, L., 2014. Along the road: aspects of causewayed enclosures in south Scandinavia and beyond. Aarhus University Press, Aarhus.

Meyer, M., Raetzel-Fabian, D., 2006. Neolithische Grabenwerke in Mitteleuropa. Ein Überblick. Journal of Neolithic Archaeology. https://doi.org/10.12766/jna.2006.20

Mischka, D., 2011. The Neolithic burial sequence at Flintbek LA3, north Germany, and its cart tracks: a precise chronology. Antiquity 85(329), 742–758.

4.1 Rastorf LA6a/c

Site name	Rastorf LA6a/c
Longitude, latitude	54.282065, 10.292363
Federal state, country	Schleswig-Holstein, Germany
Excavation campaign	1999, 2000; 2002
Excavation type	Rescue excavation
Responsible archaeologist	Dieter Stoltenberg, Archäologisches Landesamt Schleswig-Holstein
Archaeological context	House, burial site, ard marks
Analysed structures	Pits, cultural layers, hearth, megalithic tomb
Archaeological group	Funnel Beaker Culture
Archaeological periodisation	Early Neolithic to Middle Neolithic (EN II – MN I)
Absolute dating	3520 – 3100 cal BCE (according to new dates from SPP1400)
Archaeobotanists	Stefanie Klooß, Wiebke Kirleis (SPP1400); Helmut Kroll, Kiel
Archaeobotanical sampling strategy	Judgement sampling from archaeological contexts
Analysed sediment volume	c. 189 l
Find concentrations (finds/l)	6 overall; 2 for seeds and fruits, 4 nutshell fragments of hazel
Number of carpological remains	298, plus 876 nutshell fragments of hazel)

The site of Rastorf LA6a/c comprises multiple archaeological contexts. It consists of two long barrows that were erected in seven building phases. Below the burial mound, a cultural layer with numerous archaeological artefacts from a Fuchsberg assemblage dating to EN II – MN I was discovered. A house area was identified (Fig. 4.1.1 and Fig. 4.1.2). Further archaeological features at Rastorf 6a/c relate with early crop growing: criss-cross ard marks were identified outside as well as inside the house area. They seem to be younger than the house structure and cultural layer, but stratigraphically they predate the erection of the megalithic tomb. At Rastorf 6a/c, the ard marks are interpreted as originating from preparation of arable plots in a time period when the house was abandoned and the area was still cleared. For the Funnel Beaker group, it was common that megalithic tombs were erected on already open plots, be it former arable land or domestic areas; ritual ploughing is discussed as well but hard to prove (Steffens, 2009: 4).

The charred archaeobotanical material from the cultural layer comprises both domestic and wild species (Table 3). Barley, emmer, and free-threshing wheat comprise the cereal spectrum (Kroll, 2001). Nutshell fragments from hazel, unripe lime fruits, bramble, and tubers from lesser celandine show that people used resources from the still-dense Early Neolithic woodland. Postholes below the cultural layer indicate a north-east to south-west aligned two-aisled house enclosing an area of 17 × 8 m. The reconstruction shows an impressive,

Fig. 4.1.1. Reconstruction of the Rastorf LA6a/c house at the open-air museum Archäologisch-ökologisches Zentrum Albersdorf (photo: Sara Jagiolla, UFG Kiel).

massive house with a high roof ridge (Fig. 4.1.1). At a distance of 10 m north-east of the house's front face, an oval pit filled with a black, greasy substrate was excavated. It is interpreted as being contemporary with the house. It contained about 600 ceramic sherds, 700 flint flakes, animal bones and teeth from cattle and goat/sheep, and charred plant remains. The latter include linseeds, the flesh of crab apple, nutshells of hazel, and some cereal grains. The most important archaeobotanical finds are the two linseeds from the bottom of the pit, dated to the Early Neolithic (3520 – 3380 cal BCE, Table 8). They are the only evidence for the use of this species in northern Germany in the Neolithic. Recently, linseed was found in the Middle Neolithic site of Karleby, in the Falbygden area of Sweden (personal communication K.-G. Sjögren, Jan. 2018). While subfossil linseeds are known from the wetland sites e. g. of the circumalpine Neolithic lake sites (e. g. Herbig and Mayer 2011), linseeds are extremely rare finds in archaeobotanical records with charred

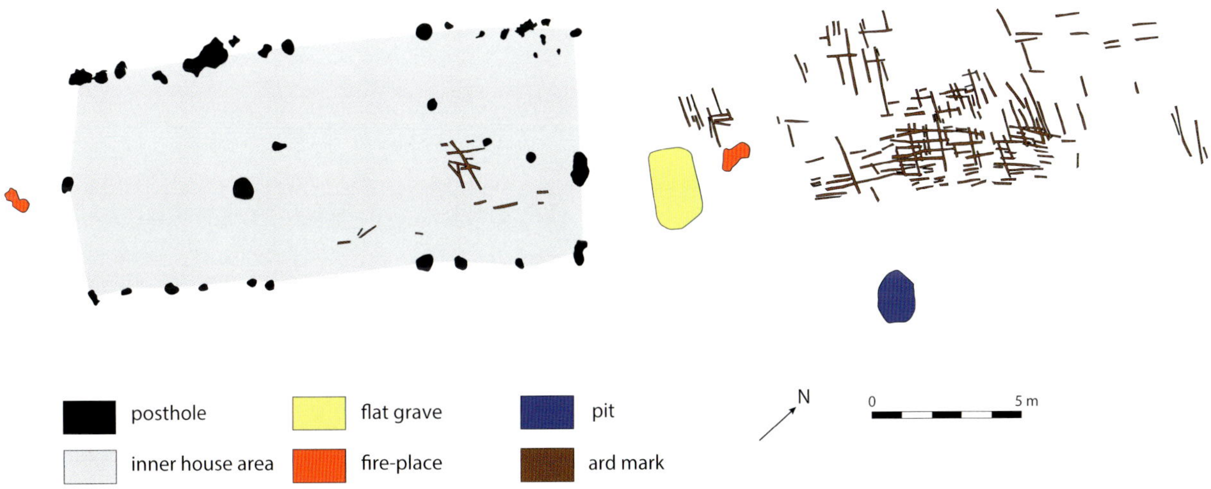

Fig. 4.1.2. Map of archaeological features, Rastorf LA6a/c (after Müller, 2017: 22).

material deposited in dry soils. They easily burn to ash because of their high oil contents. In the case of the oval pit at Rastorf LA6a/c, the presence of linseed may explain the greasy texture of the pit fill. The linseed (or flax) plant can be used in two ways. First, the seeds can be used as a valuable dietary supplement, because they contain a specific mucilage that stimulates digestion, and they can be processed to gain linseed oil. Second, the bast fibres can be used as raw material for linen cloth. Because there is no evidence for loom weights for the Funnel Beaker Culture, it is argued that textile production with weaving looms did not take place, although it is also argued that loom weights usually are connected with wool processing, because they are too heavy to be used with plant fibres (Sørensen, 2014: 29). It seems most probable that Funnel Beaker groups used the linseed plant mainly as a dietary supplement.

Fig. 4.1.3. Flowering linseed, one morning end of June (photo: Marta Dal Corso, UFG Kiel, June 2016).

Literature:

Herbig, C., Maier, U., 2011. Flax for oil or fibre? Morphometric analysis of flax seeds and new aspects of flax cultivation in Late Neolithic wetland settlements in southwest Germany, Vegetation History and Archaeobotany 20(6), 527–533.

Kroll, H., 2001. Der Mohn, die Trichterbecherkultur und das südwestliche Ostseegebiet, in: Kelm, R., (Ed.), Zurück zur Steinzeitlandschaft. Proceedings of the third conference on Archäobiologische und ökologische Forschung zur jungsteinzeitlichen Kulturlandschaft und ihrer Nutzung in Nordwestdeutschland, Albersdorf. Boyens, Heide, pp. 70–76.

Müller, J., 2017. Großsteingräber, Grabenwerke, Langhügel. Frühe Monumentalbauten Mitteleuropas. Sonderheft Archäologie in Deutschland. Theiss, Darmstadt.

Sørensen, L., 2014. From hunter to farmer in northern Europe: migration and adaptation during the Neolithic and Bronze Age. Volume 2. Wiley, Oxford.

Steffens, J., 2009. Die neolithischen Fundplätze von Rastorf, Kreis Plön. Eine Fallstudie zur Trichterbecherkultur im nördlichen Mitteleuropa am Beispiel eines Siedlungsraumes. Habelt, Bonn.

Plate 1. Rastorf LA6a/c. 1 Charred emmer grains, *Triticum dicoccon*. 2 Charred einkorn grain, *Triticum monococcum*, lateral view. 3 Charred wheat grains, *Triticum* sp.

Plate 2. Rastorf LA6a/c. 1 Tiny charred grain of bread wheat, compact variety, *Triticum aestivo-compactum*, dorsal and ventral views. 2 Charred barley grains, *Hordeum vulgare*. 3 Charred linseeds, *Linum usitatissimum,* lateral view.

Plate 3. Rastorf LA6a/c. 1 Charred nutshells of hazel, *Corylus avellana*. 2 Charred unripe hazel nuts.

Plate 4. Rastorf LA6a/c. 1 Charred crab apple, apple core and skin, *Malus sylvestris*. 2 Charred crab apple, apple flesh.

Plate 5. Rastorf LA6a/c. 1 Charred seeds of bramble, *Rubus fruticosus*. 2 Charred fruits of lime, *Tilia* sp. 3 Charred tuber of lesser celandine, *Ficaria verna*. 4 Charred scale of pine cone, *Pinus sylvestris,* top and lateral views. 5 Charred fruits of cleavers, *Galium aparine,* outer and inner views. 6 Charred seed of ribwort plantain, *Plantago lanceolata*.

4.2 Rastorf LA73

Site name	Rastorf LA73
Longitude, latitude	54.274128, 10.287677
Federal state, country	Schleswig-Holstein, Germany
Excavation campaign	2001
Excavation type	Rescue excavation
Responsible archaeologists	Hans Joachim Repkewitz, Archäologisches Landesamt Schleswig-Holstein, Klaus Hirsch, Kiel
Archaeological context	Causewayed enclosure
Analysed structures	Ditch sections
Archaeological group	Funnel Beaker Culture
Archaeological periodisation	Early Neolithic (EN Ib–EN II)
Absolute dating	3475–3371 cal BCE
Archaeobotanists	Stefanie Klooß, Wiebke Kirleis, Kiel (SPP1400)
Archaeobotanical sampling strategy	Judgement sampling from archaeological contexts
Analysed sediment volume	34 l
Find concentrations (finds/l)	1 (0.6)
Number of carpological remains	19

The causewayed enclosure site Rastorf LA73 is situated on a ridge in the eastern upper moraine hilly landscape of south-eastern Holsatia, c. 10 km south of the Kiel fjord. In the south-east, the ridge is cut by a steep terrace towards a wetland area with a tiny lake extending towards the valley of the river Schwentine. The causewayed enclosure comprises three ditches that were partly excavated. Nutshells of hazel from a ditch segment were dated to 3636–3364 cal BCE (Table 8). Thousands of archaeological artefacts were discovered in the ditches. Flakes and ceramic sherds typologically dating to EN II–MN I dominate the material, which is supplemented by some unburnt animal bones and burnt bone fragments (Steffens, 2009: 72–79).

The tiny botanical assemblage is typical for material from ditch fills of causewayed enclosures (section 3.4). It mostly comprises charred remains of wild plant species (Table 3). Crop plants are almost absent; only one cereal grain was detected. In the Neolithic, the hazel nuts and crab apple could be gathered in thickets and at the edges of woodland. The nutlet of *Carex hirta*, the tiny Poaceae fruit, the Rosaceae seed, the vetch seed, the cone spindle fragments from alder, and further vegetative remains most probably originate from the wetlands in the south-east of the enclosure site that is connected to the riverine forest along the river Schwentine. The fragments of a seed pouch of *Raphanus raphanistrum* are a more recent intrusion, since this species does not occur in the area before the Iron Age.

Literature:

Steffens, J., 2009. Die neolithischen Fundplätze von Rastorf, Kreis Plön. Eine Fallstudie zur Trichterbecherkultur im nördlichen Mitteleuropa am Beispiel eines Siedlungsraumes. Habelt, Bonn.

Plate 1. Rastorf LA73. 1 Charred seed of crab apple, *Malus sylvestris*, lateral front and back views. 2 Charred nutshell fragments of hazel, *Corylus avellana*. 3 Charred vegetative remains: culm, rhizome, or rootlet fragments. 4 Charred cone spindles of common alder, *Alnus glutinosa*. 5 Charred hairy sedge nutlet, *Carex hirta*. 6 Charred tiny grain of the grass family, Poaceae. 7 and 8 Charred seed pouch fragments of jointed charlock, *Raphanus raphanistrum*, younger intrusion.

4.3 Eisendorf LA42 and LA43

Site name	Eisendorf LA42 and Eisendorf LA43
Longitude, latitude	54.199842, 9.867911
Federal state, country	Schleswig-Holstein, Germany
Excavation campaign	2008
Excavation type	Rescue excavation
Responsible archaeologist	Steffen Haucke, Archäologisches Landesmuseum Schleswig-Holstein
Archaeological context	Non-megalithic long barrow
Analysed structures	Palisade ditch, postholes
Archaeological group	Funnel Beaker Culture
Archaeological periodisation	Early to Middle Neolithic (EN–MN)
Absolute dating	--
Archaeobotanists	Stefanie Klooß, Wiebke Kirleis, Kiel (SPP1400)
Archaeobotanical sampling strategy	Judgement sampling from archaeological contexts
Analysed sediment volume	18 l
Find concentrations (finds/l)	1
Number of carpological remains	20

A small-scale rescue excavation took place in 2008 prior to gravel exploitation in Eisendorf. Two non-megalithic long barrows and some postholes were excavated. The sides of the long barrows were accompanied by palisade ditches. In one of these non-megalithic long barrows, a chamber with a stone setting was detected, which was interpreted as a rectangular-shaped dolmen (German: Urdolmen).

The postholes and one of the palisade ditches were sampled for archaeobotanical investigations. The charred plant assemblage is very small (Table 3). It consists of four cereal grains, two of which were identified as barley, with the shape of hulled barley, several seeds of bramble and some weed seeds.

Literature: --

Plate 1. Eisendorf LA42. 1 Charred grains of barley, *Hordeum vulgare*, dorsal and ventral views. 2 Charred bramble seeds, *Rubus fruticosus*.
Eisendorf LA43. 3 Charred seed of cleavers, *Galium aparine*, outer and inner views. 4 Charred seed of false cleavers, *Galium spurium*, outer and inner views. 5 Charred seed of the Lamiaceae, labiate family, possibly woundwort, cf. *Stachys*.

4.4 Stolpe-Depenau LA14

Site name	Stolpe-Depenau LA14
Longitude, latitude	54.292857, 10.851122
Federal state, country	Schleswig-Holstein, Germany
Excavation campaign	2010
Excavation type	Rescue excavation
Responsible archaeologists	Lorenz Harten, Oliver Nakoinz, Archäolo-gisches Landesamt Schleswig-Holstein
Archaeological context	Settlement
Analysed structure	Pit
Archaeological group	Funnel Beaker Culture
Archaeological periodisation	Neolithic
Absolute dating	–
Archaeobotanists	Stefanie Klooß, Wiebke Kirleis, Kiel (SPP1400)
Archaeobotanical sampling strategy	Judgement sampling from archaeological contexts
Analysed sediment volume	18 l
Find concentrations (finds/l)	1 (0.4)
Number of carpological remains	7

The site of Stolpe-Depenau, close to the village of Nettelsee, was excavated prior to construction of the A21 motorway, in the south of Kiel. It is situated on a spur in the upper moraine plain of the Weichsel glaciation. The favourable location was revisited several times in prehistory. In addition to the Funnel Beaker site, LA14, a Younger Neolithic house structure of the Single Grave Culture was detected, LA17. Further traces date to the Metal Ages (LA16, LA18).

From the Neolithic structures at LA14, three soil samples were rescued for archaeobotanical investigations. The tiny charred archaeobotanical assemblage shows poor preservation conditions (Table 3); the depositional processes are difficult to decipher and may include intrusion of more recent material.

Literature:

Harten, L., Klooß, S., Nakoinz, O., 2011. Neolithische Siedlungsspuren unterm Weihnachtsbaum. Der Fundplatz Stolpe/Depenau, Kreis Plön, LA17. Archäologischen Nachrichten aus Schleswig-Holstein 2011, 58–61.

Plate 1. Stolpe-Depenau LA14. 1 Charred fragment of cereal grain with embryo cavity, Cerealia indet., dorsal and lateral views. 2 Charred oat grain, *Avena* sp., dorsal and ventral views. 3 Charred fruit of curly dock, *Rumex crispus*.

4.5 Bad Segeberg LA93

Site name	Bad Segeberg LA93
Longitude, latitude	54.292857, 10.851122
Federal state, country	Schleswig-Holstein, Germany
Excavation campaign	2009
Excavation type	Rescue excavation, sondage
Responsible archaeologist	Annette Guldin, Archäologisches Landesamt Schleswig-Holstein
Archaeological context	Settlement
Analysed structures	Pits, postholes
Archaeological group	Funnel Beaker Culture
Archaeological periodisation	Early Neolithic to Middle Neolithic (EN II – MN II)
Absolute dating	3500 – 3000 cal BCE
Archaeobotanists	Stefanie Klooß, Wiebke Kirleis, Kiel (SPP1400)
Archaeobotanical sampling strategy	Judgement sampling from archaeologicalcontexts
Analysed sediment volume	15 l
Find concentrations (finds/l)	1 (0.3)
Number of carpological remains	4

The site of Bad Segeberg LA93 was excavated prior to construction of the A20 motorway, on a steep, southwards-facing slope of the Kalkberg overlooking the river Trave in the valley. The main Neolithic feature is a pit with homogeneous fill (Fig. 4.5.1) that contained numerous flint artefacts and ceramic sherds of Middle Neolithic age, as well as two sherds of possible Fuchsberg periodization (EN II). From this pit only a small charred plant assemblage was recovered (Table 3). It comprises grains and threshing remains of the hulled wheats emmer and einkorn, accompanied by a nutshell of hazel and seeds of knotweed.

Fig. 4.5.1. West profile, Feature 1, Bad Segeberg LA93 (photo: ALSH, Schleswig).

Literature:

Guldin, A., 2010. Viele Funde—Kein Haus! Ausgrabung einer Grube der Jungsteinzeit (LA 93) im Trassenbereich der Bundesautobahn 20 in Bad Segeberg. Archäologische Nachrichten aus Schleswig-Holstein 2010: 38 – 40.

Plate 1. Bad Segeberg LA93. 1 Charred emmer spikelet fork, *Triticum dicoccon*. 2 Charred nutshell fragment of hazel, *Corylus avellana*. 3 Charred fruit of lady's thumb, *Persicaria maculosa*.

4.6 Bad Segeberg LA94

Site name	Bad Segeberg LA94
Longitude, latitude	53.921443, 10.296769
Federal state, country	Schleswig-Holstein, Germany
Excavation campaign	2010
Excavation type	Rescue excavation, sondage
Responsible archaeologist	Annette Guldin, Archäologisches Landesamt Schleswig-Holstein
Archaeological context	Causewayed enclosure
Analysed structures	Ditch, pits
Archaeological group	Funnel Beaker Culture
Archaeological periodisation	Early to Middle Neolithic (EN Ib/II – MN I)
Absolute dating	EN Ib/II: 3650 – 3350 cal BCE; MN I: 3350 – 3100 cal BCE
Archaeobotanists	Stefanie Klooß, Wiebke Kirleis, Kiel (SPP1400)
Archaeobotanical sampling strategy	Judgement sampling from archaeological contexts
Analysed sediment volume	1929 l
Find concentrations (finds/l)	1 (0.6)
Number of carpological remains	1081

The causewayed enclosure site of Bad Segeberg LA94, in the vicinity of the site of Bad Segeberg LA93, was excavated prior to construction of the A20 motorway. The enclosure area of 10 000 m^2 is located on a spur on the bank of the Trave river. In the north, it is closed off by two ditch segments, each of 40 m length, 10 m width, and 2.5 m depth. The segments delineate a passage into the enclosure site (Fig. 4.6.1). While the inner part of the enclosure does not show features related to any construction, it does show a cultural layer with numerous ceramic sherds and stone artefacts. The ditch is filled with about 11 000 ceramic sherds and 17 000 stone artefacts, including flint with sickle gloss. Two phases can be distinguished (Guldin, 2011).

The archaeobotanical assemblage from the Early Neolithic (EN Ib/II) is small (Tables 3 and 8). Some cereal grains and threshing remains of emmer, einkorn, and barley are present. Remains of gathered plants, such as hazel, blackberry, and crab apple, supplement the cereals. The number of finds of root tubers of lesser celandine is quite high. These starch-rich tubers may have been an additional food source (Klooß et al., 2016). Among the weed species, the small Poaceae, namely, timothy grass, *Phleum pratense*, are dominant. The remaining weeds include goosegrass, white goosefoot, false cleavers, and black bindweed. The plant assemblage from the Middle Neolithic (MN I) shows higher numbers of finds. Emmer, einkorn, and barley grains provide evidence for cereal consumption. Numerous emmer glume bases show that dehusking took place at the site. Again gathered plants play a prominent role; hazel, crab apple, hawthorn, blackberry, raspberry, and root tubers of lesser celandine were also

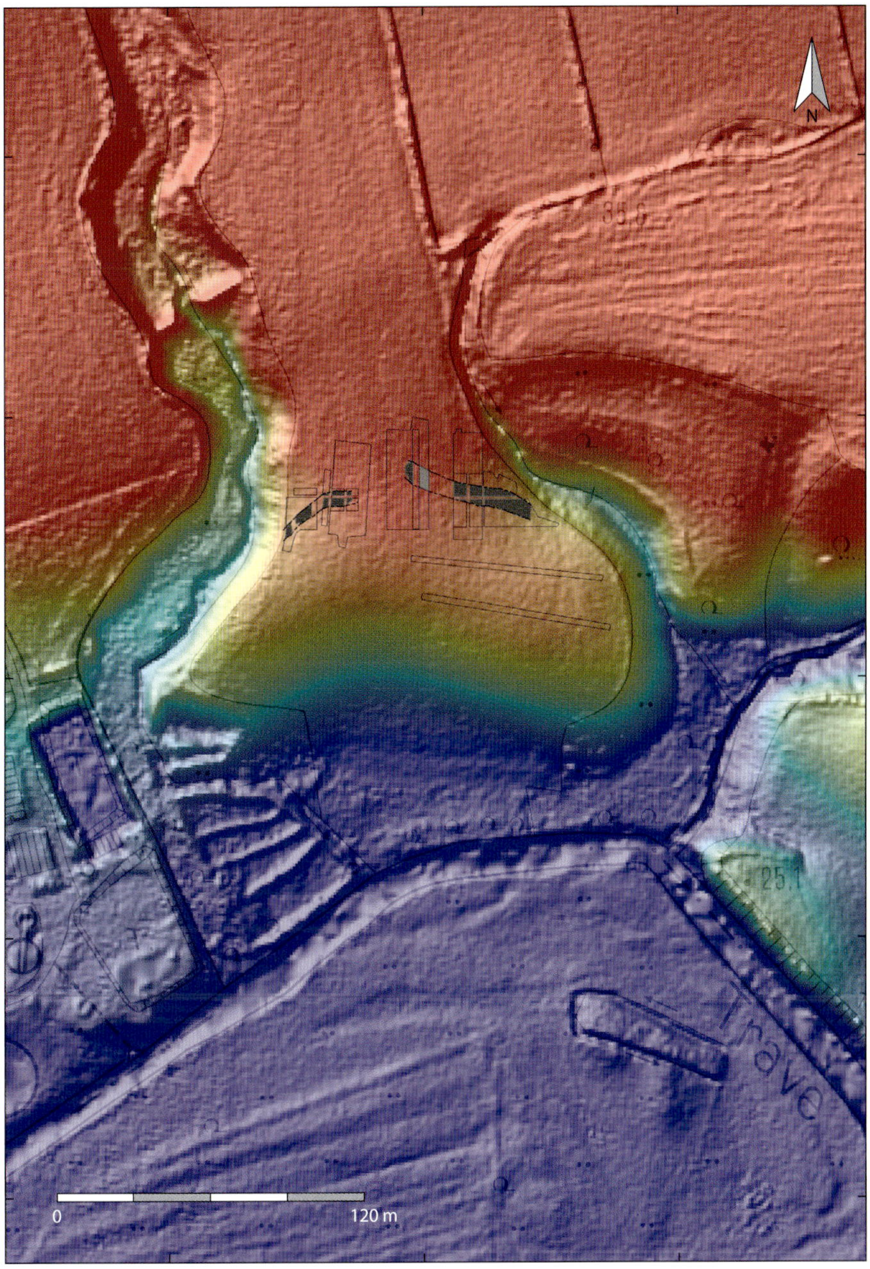

Fig. 4.6.1. Lidar scan of the causewayed enclosure of Bad Segeberg LA94, on a spur above the river Trave, showing excavation trenches and the single ditch in the north that closes off the area. (Map: LVerm Geo SH, processed by Annette Guldin, ALSH, Schleswig).

used as food plants. The weed spectrum is dominated by Poaceae, with again timothy grass as main component. Together with ribwort plantain, it shows ongoing enlargement of the arable fields accompanying an extensification of crop growing.

Literature:

Guldin, A., 2011. Eine Straße erzählt Geschichte(n)... Ein neu entdecktes Erdwerk der Jungsteinzeit im Trassenbereich der geplanten Autobahn 20 bei Bad Segeberg. Archäologische Nachrichten aus Schleswig-Holstein 2011: 33 – 35.

Klooβ, S., Fischer, E., Out, W., Kirleis, W., 2016. Charred root tubers of lesser celandine (Ficaria verna HUDS.) in plant macro remain assemblages from northern, central and western Europe. Quaternary International 404, 25 – 42.

Plate 1. Bad Segeberg LA94. 1 Charred cereal grains, Cerealia indet. 2 Charred grain of emmer, *Triticum dicoccon*, dorsal and lateral views. 3 Charred grain of emmer, dorsal, lateral, and ventral views. 4 Charred spikelet fork of hulled wheat, einkorn/emmer, *T. monococcum/dicoccon*.

Plate 2. Bad Segeberg LA94. 1 – 3 Charred grains of barley, *Hordeum vulgare*, dorsal and ventral views. 4 Charred nutshell fragment of hazel, *Corylus avellana*. 5 and 6 Charred root tubers of lesser celandine, *Ficaria verna*. 7 Charred tuber of tuber oat grass, *Arrhenatherum elatius* var. *bulbosum*. 8 Charred seed of raspberry, *Rubus idaeus*.

Plate 3. Bad Segeberg LA94. 1 Charred seed of cleavers, *Galium aparine*, outer view. 2 Charred seeds of white goosefoot, *Chenopodium album*. 3 and 4 Charred fruits of timothy grass, *Phleum pratense*.

4.7 Bosau LA242

Site name	Bosau LA242, Vorderste Waade
Longitude, latitude	53.922839, 10.293981
Federal state, country	Schleswig-Holstein, Germany
Excavation campaign	1978
Excavation type	Rescue excavation (SFB 17)
Responsible archaeologist	Herrmann Hinz, Kiel
Archaeological context	Domestic site
Analysed structures	Pit
Archaeological group	Early Dagger groups
Archaeological periodisation	Late Neolithic
Absolute dating	2154–1879 cal BCE (according to conventional radiocarbon data)
Archaeobotanist	Helmut Kroll, Kiel
Archaeobotanical sampling strategy	Judgement sampling from archaeological contexts
Analysed sediment volume	c. 60 l
Find concentrations (finds/l)	103
Number of carpological remains	6150

The site of Bosau LA242, Vorderste Waade, is situated on a dam-like spur in the northern portion of the Bischofssee, a bay on the eastern shoreline of the Großer Plöner See separated by spits from the main basin (Clausen, 1985). During excavations in 1978, in trench IX, a single pit was identified that contained a fill mixed with numerous cereal grains. Conventional radiocarbon dating of some of the cereal grains yielded a Late Neolithic date (Table 8). Only very few sites are known from the Late Neolithic period in the north. The archaeological groups prevailing at that time are named Dagger groups because of their common artefacts, stone daggers (Kühn, 1979).

The archaeobotanical investigation of the agglomeration of cereal grains was carried out by Kroll (1980). The plant assemblage is dominated by grains of free-threshing wheat, accompanied by free-threshing barley. Single emmer grains, hazel nutshell fragments, and dogwood seeds supplement the mass find.

As noted e.g. by Kirleis and Fischer (2014), free-threshing wheat is one of the four main cereals of the Early Neolithic Funnel Beaker groups. From the Middle Neolithic onwards, free-threshing wheat is present only as minor component of the crop assemblages (Fig. 4.7.1). These observations are built upon representative data, including only those sites that fulfil the requirement of at least three investigated features/cultural layers providing a minimum of 25 grains per site. Mass finds, such as the one from Bosau LA242, were excluded for this overview (Kirleis and Fischer, 2014). The mass find of free-threshing wheat from Bosau LA242 is exceptional, in particular considering its Late Neolithic date. It is an assemblage of naked grains only (Table 3),

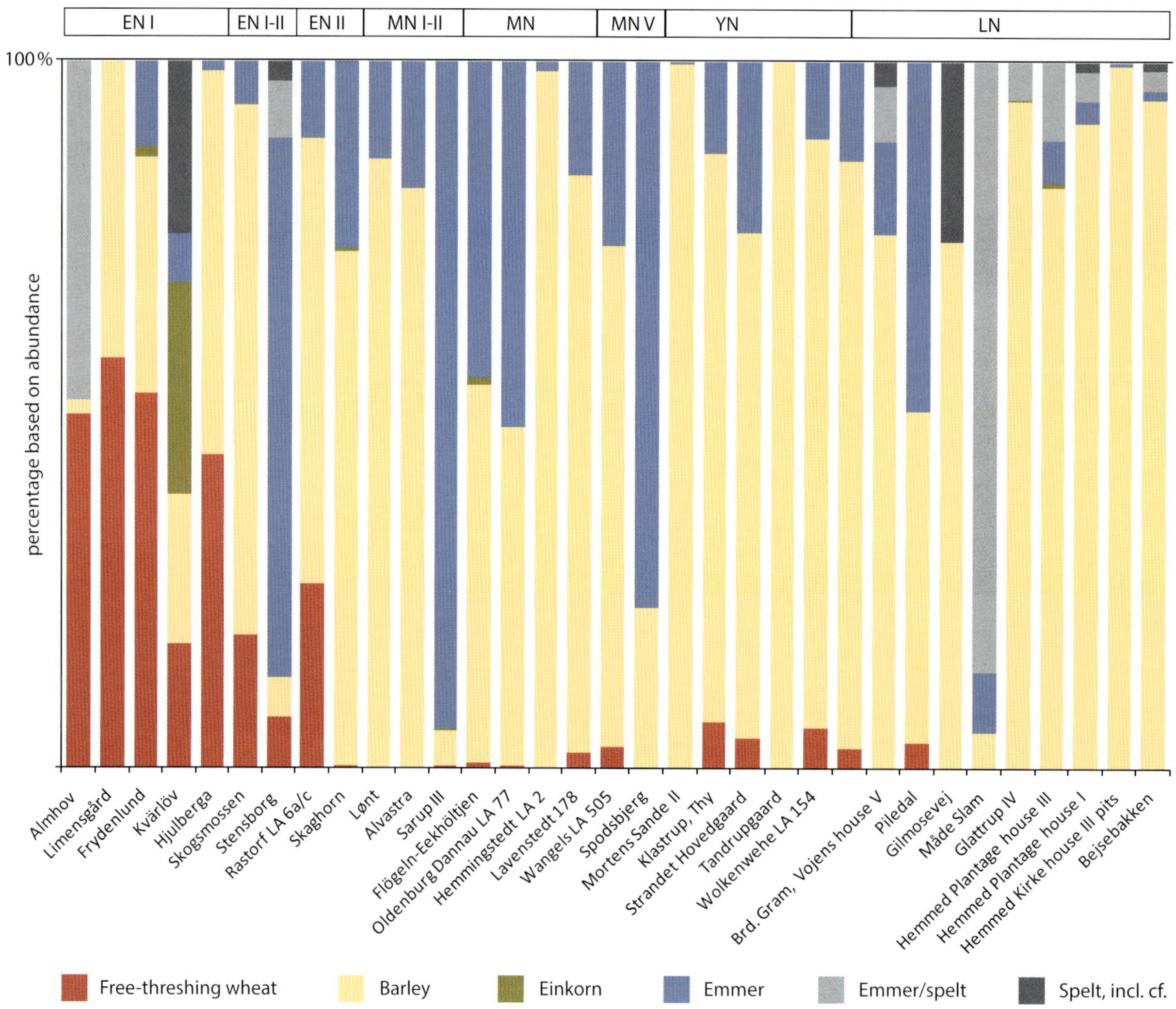

Fig. 4.7.1. Comparison of representative Neolithic cereal spectra in the area of the Funnel Beaker North Group (after Kirleis/Fischer 2014: Fig. 6).

without any hint of threshing remains. Since a differentiation between the tetraploid and the hexaploid chromosome constitution of wheat relies on finds of rachis segments with specific morphology, it cannot be clarified if this mass find consists of grains indicating the tetraploid *Triticum durum/turgidum* or the hexaploid *Triticum aestivum*. Funnel Beaker groups seem to prefer the tetraploid free-threshing wheat that was identified from the Early Neolithic sites of Albersdorf LA68, Dieksknöll (section 3.4), and Frydenlund (section 7.2) and, through re-investigation of the archived material, for the MN V site of Wangels LA505 (section 5.4). Further archaeobotanical studies are needed to prove whether, towards the Late Neolithic, a change in the crop assemblage takes place and whether the Funnel Beaker groups had different crop choices than the Early Dagger groups. Unfortunately, archaeological traces of Late Neolithic settlements are extremely rare.

Literature:

Clausen, I., 1985. Neolithische Fundplätze am Großen Plöner See. Offa 42, 113–186.

Kirleis, W., Fischer, E., 2014. Neolithic cultivation of tetraploid free threshing wheat in Denmark and northern Germany: implications for crop diversity and societal dynamics of the Funnel Beaker Culture. Vegetation History and Archaeobotany 23, Supplement 1, 81–96.

Kroll, H., 1981. Einige vorgeschichtliche Vorratsfunde von Kulturpflanzen aus Norddeutschland. Offa 37, 372–383.

Kühn, H.J., 1979. Das Spätneolithikum in Schleswig-Holstein. Offa-Bücher 40. Wachholtz, Neumünster.

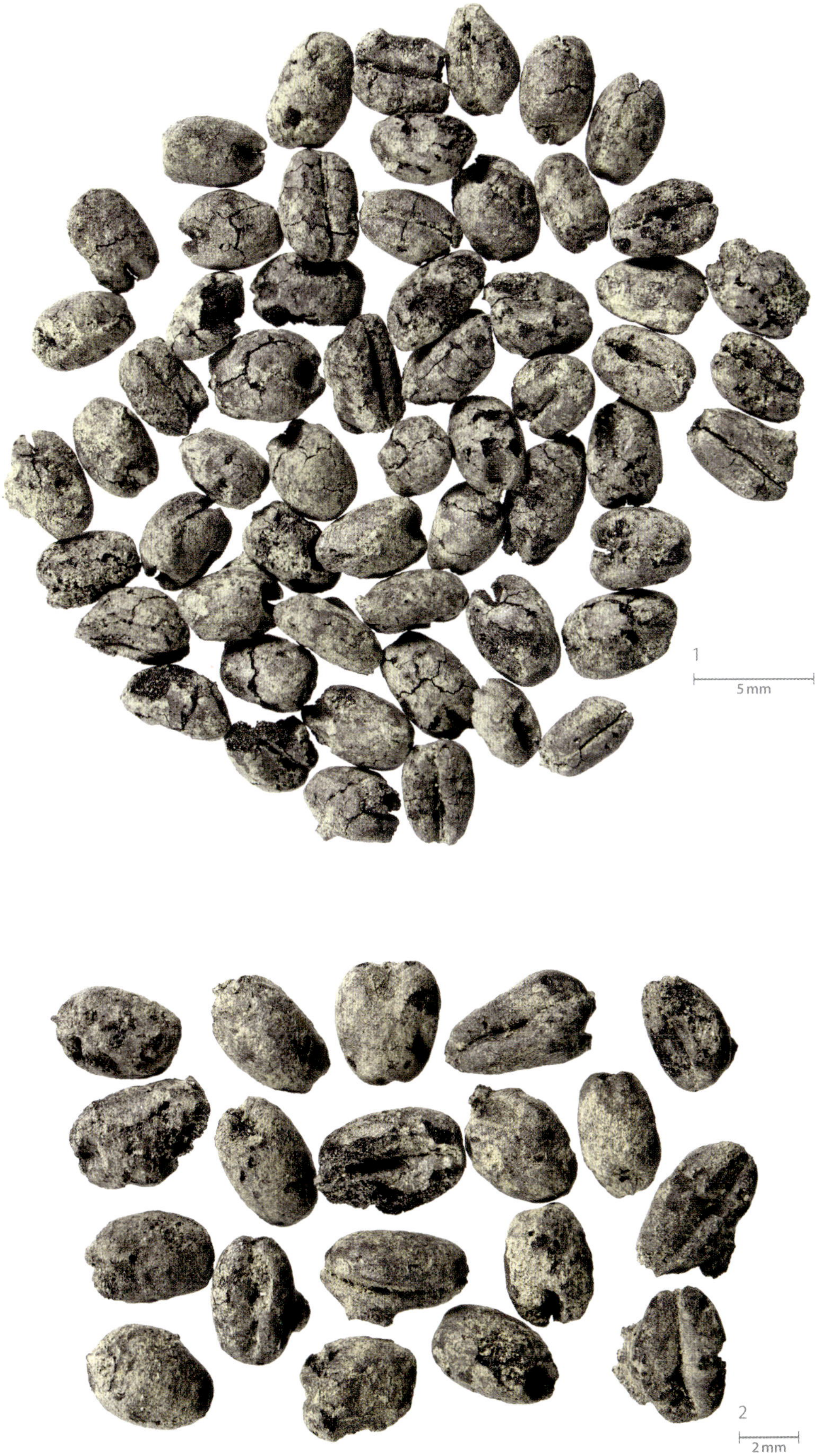

Plate 1. Bosau LA242. 1 and 2 Mass find of charred grains of free-threshing wheat, *Triticum aestivum* s. l.

Plate 2. Bosau LA242. 1– 6 Single charred grains of free-threshing wheat, *Triticum aestivum* s. l., dorsal and ventral views.

4.8 Flintbek LA137 and Flintbek LA167

Site name	Flintbek LA137 and Flintbek LA167
Longitude, latitude	LA 137: 54.253944, 10.096445 LA 167: 54.253507, 10.096173
Federal state, country	Schleswig-Holstein, Germany
Excavation campaigns	LA137: 1987; LA167: 1993 – 94
Excavation type	Rescue excavation
Responsible archaeologist	Dieter Stoltenberg, Archäologisches Landesmuseum Schleswig-Holstein
Archaeological context	Burial site
Analysed structures	Megalithic long barrows
Archaeological group	Funnel Beaker Culture
Archaeological periodisation	LA 137: Early Neolithic (EN II); LA 167: Early Neolithic (EN I)
Absolute dating	LA 137: 3465 – 3370 cal BCE; LA 167: 3900–3535 cal BCE
Archaeobotanist	Wiebke Kirleis, Kiel (SPP1400)
Archaeobotanical sampling strategy	Ceramic sherds with imprints of plant remains
Analysed sediment volume	n.a.; seven sherds were analysed
Find concentrations (finds/l)	–
Number of carpological remains	–

The Neolithic to Early Bronze Age settlement cluster of Flintbek is situated 8 km south-west of Kiel. Between 1976 and 1996, a cemetery containing 88 Neolithic and Bronze Age burials was intensively excavated by the state office of Schleswig-Holstein in the context of recent plough damage (Fig. 4.8.1). The archaeological Neolithic research focused on the ritual Funnel Beaker sites, such as non-megalithic long barrows and megalithic tombs (Zich 1992/1993; Mischka 2011 a – c). Flintbek is famous for the oldest dated cart tracks and ard marks from northern Germany (Fig. 4.8.2), found at the site LA3 and dated to between 3420 and 3385 cal BCE (Mischka 2011 a).

Archaeobotanical investigations dealt with plant imprints from ceramic sherds that were deposited in two Early Neolithic long barrows. Positive casts were produced from the negative imprints with the aid of flasking plaster from a dental laboratory (Panasil contact plus X-light). Out of a total of seven sherds with imprints from the sites Flintbek LA137 and LA167, three imprints from sherds of three ceramic vessels could be identified as originating from plant remains. While vessel 35 from LA167 only shows an imprint of charcoal, imprints from the sherds of vessels 10 and 14 from LA137 derive from cereal remains, representing an imprint from an emmer spikelet and an emmer spikelet fork with glumes. Emmer is one of the main cereals in Middle Neolithic Funnel Beaker contexts.

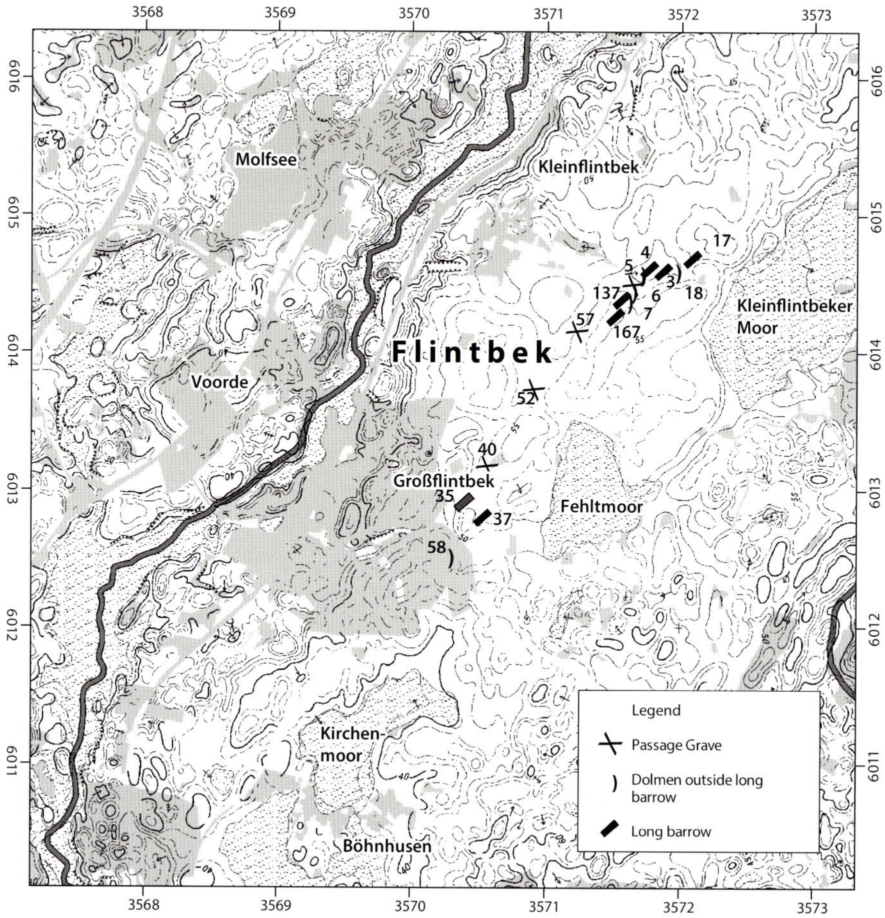

Fig. 4.8.1. Map of the Flintbek micro-region in Schleswig-Holstein, showing sites excavated 1977–1996 (after Zich, 2005: Fig. 1).

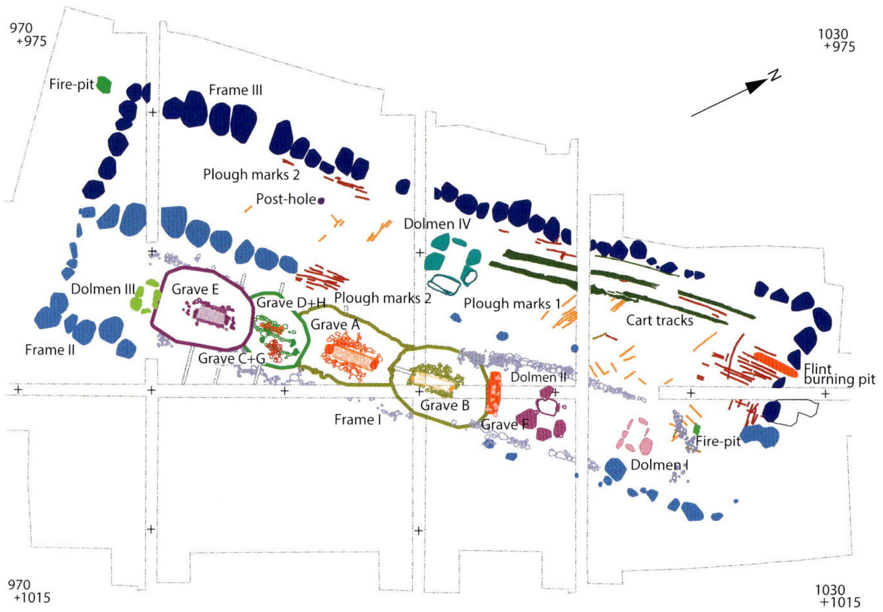

Fig. 4.8.2. Ard marks and cart tracks below the long barrow Flintbek LA3, dated to between 3420 and 3385 cal BCE. Mapped is the final stage of the long barrow, showing eight non-megalithic graves (A–H), four megalithic dolmen chambers (I–IV), three stone frames, three firepits, a posthole, two differently orientated sets of plough marks, and the famous cart tracks in the north-eastern part (after Mischka 2011 b).

Literature:

Mischka, D., 2011 a. The Neolithic burial sequence at Flintbek LA3, north Germany, and its cart tracks: a precise chronology. Antiquity, 85(329), 742–758.

Mischka, D., 2011 b. Flintbek LA3, biography of monument, in: Furholt, M., Lüth, F., Müller, J. (Eds.), Megaliths and identities: early monuments and Neolithic societies from the Atlantic to the Baltic. Habelt, Bonn, pp. 67–94.

Mischka, D., 2011 c. Das Neolithikum in Flintbek, Kr. Rendsburg-Eckernförde, Schleswig-Holstein. Eine feinchronologische Studie zur Besiedlungsgeschichte einer Siedlungskammer anhand von Gräbern. Habilitation thesis, Christian-Albrechts-University, Kiel.

Zich, B., 1992/1993. Die Ausgrabungen chronisch gefährdeter Hügelgräber der Stein- und Bronzezeit in Flintbek, Kreis Rendsburg-Eckernförde. Ein Vorbericht. Offa 49/50, 15–31.

Plate 1. Flintbek LA137. 1 and 2 Imprint of an emmer spikelet, *Triticum dicoccon*, in a sherd of vessel 14, negative imprint and positive cast.
Flintbek LA167. 3 Imprint of an emmer spikelet fork with glumes, in a sherd of vessel 10, negative imprint and positive cast.

5 North-eastern Holsatia, Schleswig-Holstein: The Oldenburger Graben in the upper moraine landscape

A favourable ecological niche in north-eastern Holsatia is the Oldenburger Graben, a wetland area in the Wagrien peninsula (Fig. 5.1). It is situated in the eastern hilly landscape that was formed during the Weichsel glaciation, in the period c. 115 000 – 11 500 BCE. The upper moraine landscape shows a dynamic morphology with eroded summits, slopes, and depressions as well as areas with slight relief. The highest elevation is +167 m above mean sea level (msl), at the top of the Bungsberg. The main soil substratum is boulder clay, and the topsoil is luvisol. The approximately 23 km long and 3 km wide wetland area known as the Oldenburger Graben cuts the Wagrien peninsula in two. It lies below msl and consists of peaty areas and an inland lake that is increasingly

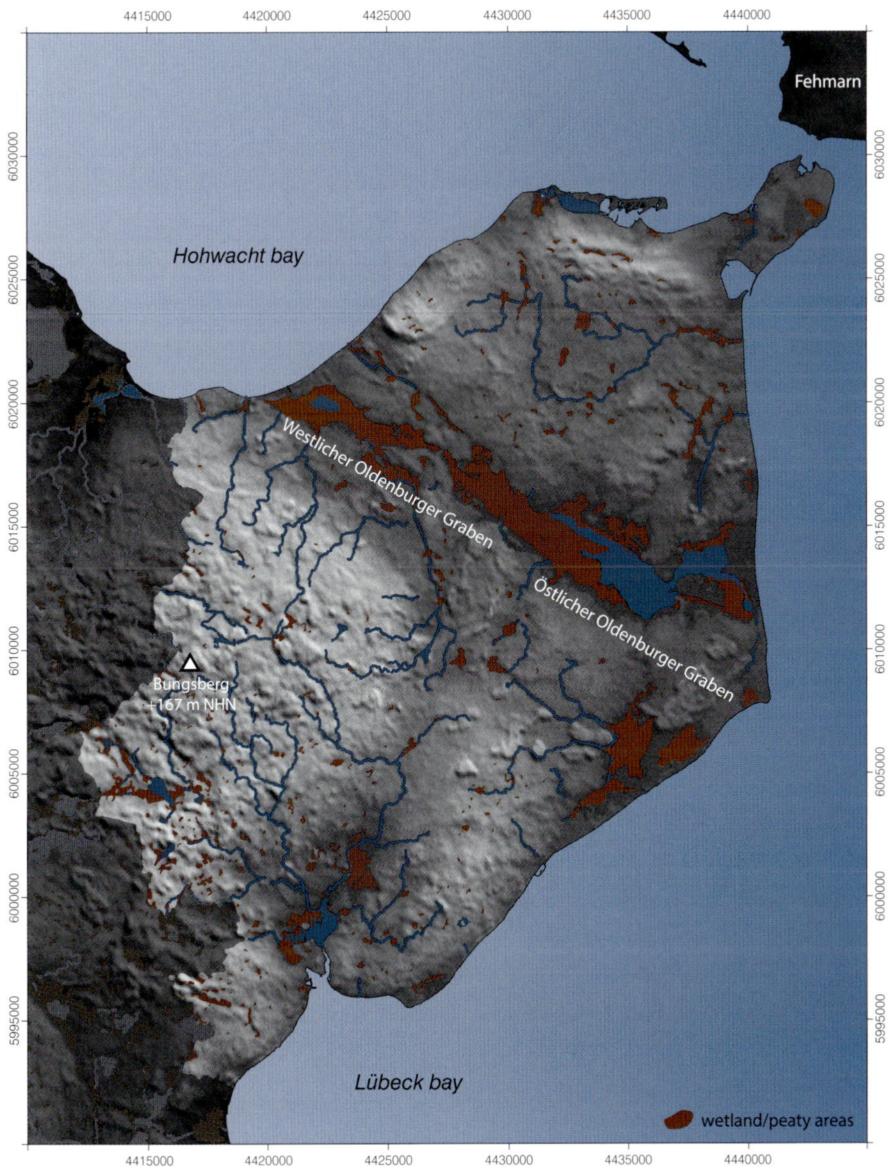

Fig. 5.1. The Wagrien peninsula, in north-eastern Holsatia, Schleswig-Holstein, geomorphologically divided into a southern and a northern part by the Oldenburger Graben; NHN refers to elevation above mean sea level (after Brozio, 2016: Fig. 5).

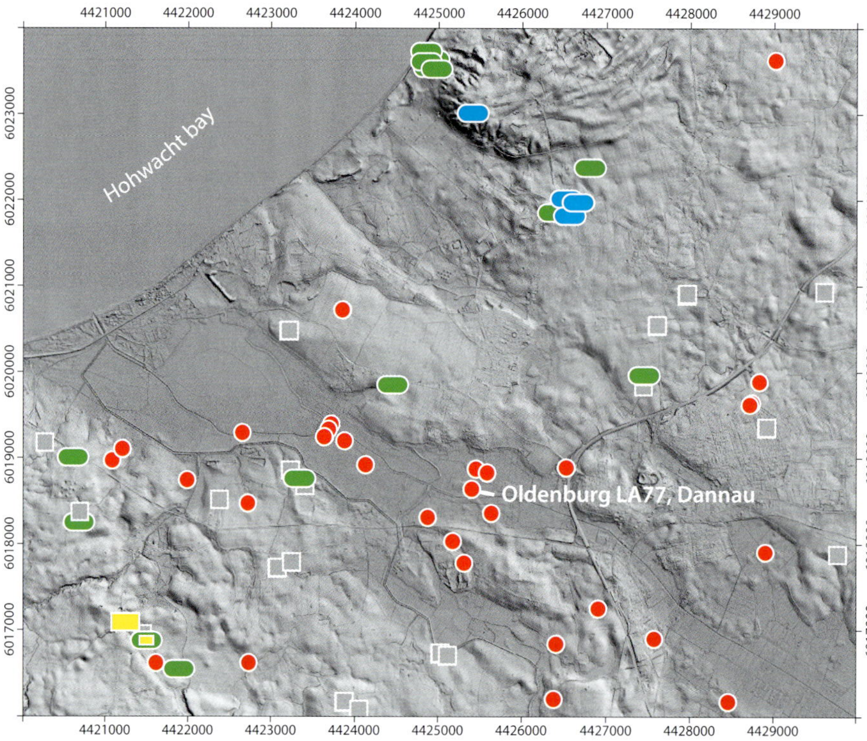

Fig. 5.2. The Neolithic landscape of the western Oldenburger Graben, showing the distribution of domestic and ritual sites (after Brozio, 2016: Fig. 171, ©LVermGeo-SH).

silting up. Levees have long since stopped the Baltic from entering the inland lowland areas, but before the first levees were constructed, sea-level changes recurrently led to landscape transformations in the area. In the late Mesolithic and the Early Neolithic, the Westlicher Oldenburger Graben was still connected with the Baltic in the Hohwacht bay. Only from 3000 BCE onwards was the area cut off from the Baltic and was the marine influence reduced due to the establishment and expansion of levees. The development of a lagoon area with brackish water conditions lasted until approx. 2900 BCE, when the conditions changed to limnic (Jakobsen, 2004; Venus, 2004). The diversity of the natural environment offered favourable conditions for human activity in the past. The terrestrial, freshwater aquatic, and marine environments ensure the availability of a wide array of resources. The area of the Oldenburger Graben in particular was recurrently revisited from prehistoric to medieval times. In the Neolithic, sandy islands in the lagoon were preferred settlement locations. Intensive archaeological investigations were carried out in the western part of the Oldenburger Graben in the frame of the priority program SPP1400 'Early Monumentality and Social Differentiation' (Fig. 5.2). The Funnel Beaker site of Oldenburg LA77, Dannau (section 5.1) is one of the oldest villages of Schleswig-Holstein, consisting in the course of time of 18 to 36 houses plus subsidiary structures (Brozio, 2016). Based on these data, a diachronic development of the settlement patterns was reconstructed, starting in the early Neolithic with single farmsteads and shifting in the Middle Neolithic towards population agglomeration in villages (Müller, 2017).

Literature:

Brozio, J.P., 2016. Megalithanlagen und Siedlungsmuster im trichterbecherzeitlichen Ostholstein. Habelt, Bonn.

Jakobsen, O., 2004. Die Grube-Wesseker Niederung (Oldenburger Graben, Ostholstein). Quartärgeologische und geoarchäologische Untersuchungen zur Landschaftsgeschichte vor dem Hintergrund des anhaltenden postglazialen Meeresspiegelanstiegs. Ph.D. thesis, Universität Kiel, Kiel.

Müller, J., 2017. Großsteingräber, Grabenwerke, Langhügel. Frühe Monumentalbauten Mitteleuropas. Theiss, Darmstadt.

Venus, J., 2004. Pollenanalytische Untersuchungen zur Vegetations- und Siedlungsgeschichte Ostwagriens und der Insel Fehmarn. Offa 82, 31–94.

5.1 Oldenburg LA77, Dannau

Site name	Oldenburg LA77, Dannau
Longitude, latitude	54.292857, 10.851122
Federal state, country	Schleswig-Holstein, Germany
Excavation campaign	2009–2012
Excavation type	Research excavation (SPP1400)
Responsible archaeologists	Jan Piet Brozio, Johannes Müller, Kiel
Archaeological context	Settlement
Analysed structures	Pits, postholes, cultural layers, burials
Archaeological group	Funnel Beaker Culture
Archaeological periodisation	Middle Neolithic (MN I–MN IV)
Absolute dating	c. 3300–3050 cal BCE
Archaeobotanists	Stefanie Klooß, Henrike Röhrs, Wiebke Kirleis, Kiel (SPP1400)
Archaeobotanical sampling strategy	Every feature sampled plus every fifth quadrant of the cultural layers
Analysed sediment volume	c. 3700 l
Find concentrations (finds/l)	c. 11 for the charred remains, 10 for the waterlogged remains
Number of carpological remains	40646 (charred); est. 35539 (waterlogged, extrapolated)

Fig. 5.1.1. Charred plant assemblage from the well feature, Oldenburg LA77, Dannau (photo: Wiebke Kirleis, UFG Kiel).

The Middle Neolithic settlement site (Table 8) of Oldenburg LA77, Dannau is situated on a sandy island in the wetland area called Oldenburger Graben in eastern Holsatia, which was cut off from the Baltic during the Neolithic. The settlement thus was surrounded by a water body with marine to brackish

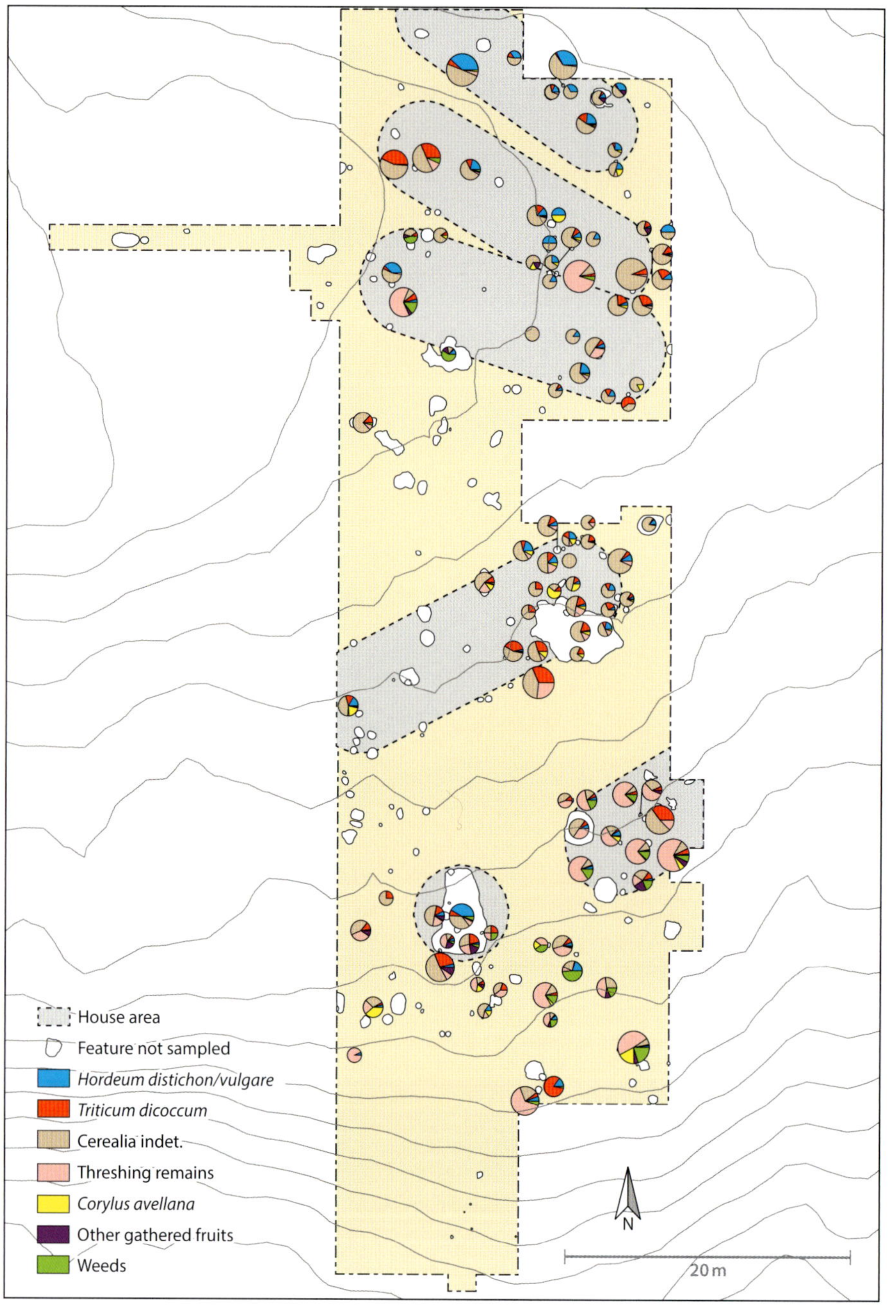

Fig. 5.1.2. House outlines and find concentrations of plant remains per sampled feature, Oldenburg LA77, Dannau; the dune top lies to the north, the wetland to the south (graphic: Stefanie Klooß, Jan Piet Brozio, Agnes Heitmann, UFG Kiel).

conditions. Oldenburg LA77, Dannau is one of the oldest villages of Schleswig-Holstein and consisted in the course of time of 18 to 36 houses plus subsidiary structures. The settlement area is defined by a cultural layer covering c. 1.35 ha. Excavations concentrated on a transect in the eastern part of the site, extending from the uppermost part of the dune southwards, towards the former shoreline (Brozio, 2016). Pits and postholes are the most common features. Single burials and deposition of human bones show that the daily routines were intertwined with ritual activity. Exceptional are two well features with clearly stratified secondary infill (Brozio et al., 2014). Numerous flint artefacts indicate local stone tool production. The ceramic inventory is highly fragmented and relates to for daily domestic activities. Food preparation is indicated by finds of quern stones that most probably were in use for cereal processing.

Oldenburg LA77, Dannau yielded by far the most extensive and multi-faceted archaeobotanical dataset for the Funnel Beaker groups anywhere in northern Germany (Table 4). The crop plant assemblage is dominated by emmer and free-threshing barley. Einkorn and bread wheat/durum wheat were of minor importance. Glume bases of emmer and einkorn show that the hulled wheats were stored in the form of spikelets, while rachis segments from barley hint at storage in the form of intact ears for this species. There is evidence for the use of linseed and opium poppy, however sparse (based on single seed finds for each species). Pulses did not form part of the crop spectrum. The spatial distribution of the cereal remains shows that the grains dominate on the dune top and relate to the house plans. In contrast, the threshing remains cluster mainly towards the wetland area in the south. They occur outside house structures and are associated with one structure that possibly was dedicated to cereal processing activities. Thus, food production was located in the centre of the settlement, while cereal processing and refuse disposal mainly took place at the settlement's border, towards the shoreline (Fig. 5.1.2). Numerous charred weed finds (>3700 seeds and fruits) were identified from Oldenburg LA77, Dannau. The most important species numerically are *Chenopodium album, Phleum pratense, Poa annua, Bromus arvensis/secalinus, Solanum nigrum, Fallopia convolvulus, Persicaria maculosa, Plantago lanceolata, Trifolium* sp., and *Vicia* sp. (Table 4). Evidence for the perennial species *Phleum pratense* and *Plantago lanceolata* as arable weeds, in particular, indicates low labour input in weeding. Thus, extensive cereal cultivation can be assumed, whereby the arable plots were presumably prepared for sowing by use of the ard. Most probably the arable plots were located in the vicinity of the settlement. A fuzzy-based vegetation reconstruction suggests that the entire area along the Oldenburger Graben was available for human appropriation, as mirrored by the varying extent of light forest and agricultural land (Knitter et al., 2019). The plant food spectrum was supplemented by hazel nuts and crab apple. Numerous charred finds of crab apple occurred in one of the well-features. Many of the crab apples had been cut in half when they were still fresh, as a way of preparing them for drying before storage as staple fruit. Most probably the apples were accidently charred once dried and then disposed of together with other settlement refuse (Brozio et al., 2014). The high number of crab apple finds suggests gathering in the light woodlands, where stands of crab apple were increasing due to the activities of the Neolithic people and their domestic animals. However, since twigs and bark of this species, as well as the fruits, are favoured fodder for domestic and wild animals, it is highly probable that the light woodlands were being managed to ensure high crab apple yields for human consumption. It is even possible that crab apple groves were established in the settlement's vicinity and enclosed by hedges or fences as early as the Neolithic (Kirleis, 2018). Harvesting crab apples with a group of students in 2018, a year with favourable weather conditions for apple growth, yielded 50 to 70 kg crab apples per tree of about 10 m height growing in cramped woodland edges. This high number of fruits shows that crab apple in the Neolithic must have been a welcome and important source for an additional staple food.

Literature:

Brozio, J.P., 2016. Megalithanlagen und Siedlungsmuster im trichterbecherzeitlichen Ostholstein. Habelt, Bonn.

Brozio, J.P., Dörfler, W., Feeser, I., Kirleis, W., Klooß, S., Müller, J., 2014. A Middle Neolithic well from northern Germany: a precise source to reconstruct water supply management, subsistence economy, and deposition practices. Journal of Archaeological Science 51, 135–153.

Kirleis, W., 2018. The cultural significance of plants, in: Haug, A., Käppel, L., Müller, J. (Eds.), Past landscapes: the dynamics of interaction between society, landscape, and culture. Sidestone, Leiden, pp. 169–182.

Knitter, D., Brozio, J.P., Dörfler, W., Duttmann, R., Feeser, I., Hamer, W., Kirleis, W., Müller, J., Nakoinz, O., 2019 (accepted). Transforming landscapes: land-use patterns of environmental borderlands. The Holocene.

Plate 1. Oldenburg LA77, Dannau. 1 Charred assemblage of barley grains, *Hordeum vulgare*. 2 Charred grain of free-threshing barley, *Hordeum vulgare nudum*, dorsal and lateral views. 3 Charred emmer remains, *Triticum dicoccon* (photos: Stefanie Klooß, UFG Kiel).

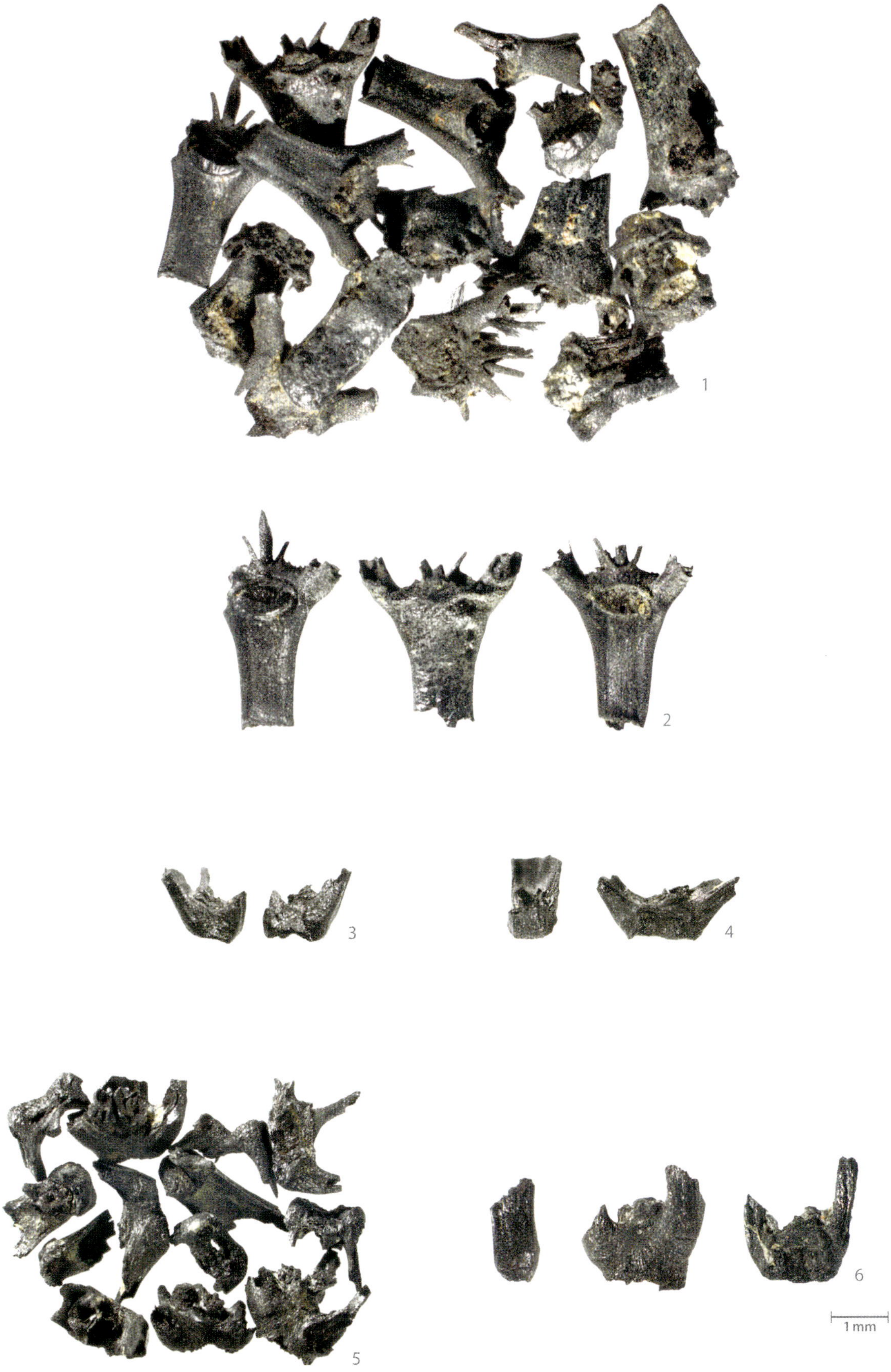

Plate 2. Oldenburg LA77, Dannau. 1 and 2 Charred rachis segments of barley, *Hordeum vulgare*. 3 and 4 Charred uppermost spikelet fork of emmer, *Triticum dicoccon*. 5 and 6 Charred emmer spikelet forks.

Plate 3. Oldenburg LA77, Dannau. 1–4 Charred crab apples, *Malus sylvestris.* 5 Small Poaceae, timothy grass/annual meadow grass, *Phleum pratense/Poa annua* (photos: Sara Jagiolla (1–3), Stefanie Klooß (5), UFG Kiel).

5.2 Oldenburg LA191

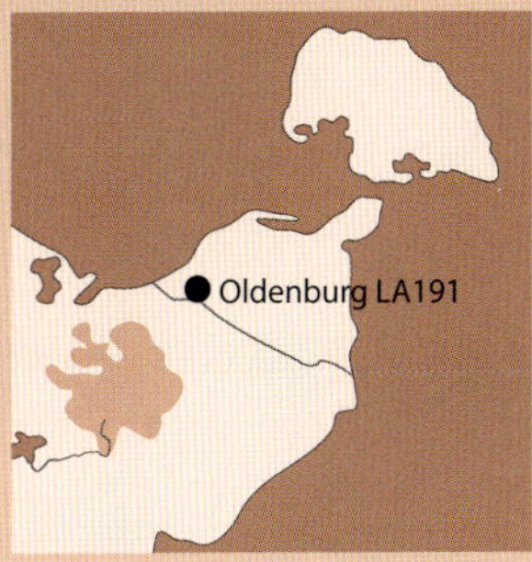

Site name	Oldenburg LA191
Longitude, latitude	54.292857, 10.851122
Federal state, country	Schleswig-Holstein, Germany
Excavation campaign	1979–1980; 1996–1997
Excavation type	Research excavation
Responsible archaeologist	Jürgen Hoika, Museum für Archäologie Schloss Gottorf
Archaeological context	Settlement
Analysed structures	Pits, postholes, cultural layers
Archaeological group	Funnel Beaker Culture
Archaeological periodisation	Early to Middle Neolithic (EN II–MN IV)
Absolute dating	c. 3500–2900 cal BCE
Archaeobotanists	Helmut Kroll, Stefanie Klooß, Elske Fischer, Dragana Filipović, Kiel (SPP1400, SFB1266)
Archaeobotanical sampling strategy	Judgement sampling from archaeological contexts
Analysed sediment volume	c. 210 l
Find concentrations (finds/l)	c. 45
Number of carpological remains	c. 9423

Similar to Oldenburg LA77, Dannau, the settlement site of Oldenburg LA191 is situated on a sandy island in the wetland area called Oldenburger Graben, in eastern Holsatia (Fig. 5.1; Hoika, 1981, 1987). The ceramic inventory and new radiocarbon data (Table 8) allow for a periodisation to the phases MN I–IV.

Archaeobotanical investigations were carried out on both waterlogged and dry soil samples (Kroll, 1981; Filipović et al. 2019). Insight into the plant economy derives from the charred assemblages. Free-threshing barley and emmer are the dominant species (Table 4). Einkorn is present as well. Emmer glume bases occur on a regular basis, while only four rachis segments of barley are present. Thus, de-husking of emmer spikelets took place in the settlement, and the free-threshing barley was presumably stored as grains. Gathered fruits supplemented the diet, as is shown by charred seeds of crab apple and raspberry *Rubus idaeus* and nutshells of hazel. Charred weed finds mainly consist of small Poaceae fruits. Additional information on the natural environment derives from subfossil plant remains.

Literature:

Filipović, D., Brozio, J.P., Kaczmarek, J., Müller, J., Kirleis, W., 2019 submitted. Food transformed? Taphonomical investigation into a potentially symbolic role of crops at two Neolithic settlements in northern Germany, Prähistorische Zeitschrift.

Hoika, J., 1981. Jungsteinzeitliche Siedlungsreste aus Oldenburg-Dannau, Kreis Ostholstein. Offa 38, 5–83.

Hoika, J., 1987. Das Mittelneolithikum zur Zeit der Trichterbecherkultur in Nordostholstein. Untersuchungen zur Archäologie und Landschaftsgeschichte. Offa-Bücher 61. Wachholtz, Neumünster.

Kroll, H., 1981. Mittelneolithisches Getreide aus Dannau. Offa 38, 85–90.

Plate 1. Oldenburg LA191. 1 Charred emmer and barley mass find, *Triticum dicoccon* and *Hordeum vulgare*. 2 Charred emmer/barley assemblage, zoomed in.

Plate 2. Oldenburg LA191. 1 Charred barley grains, *Hordeum vulgare*. 2 – 4 Charred barley grains, dorsal and ventral views. 5 Charred emmer spikelet forks, *Triticum dicoccon*.

Plate 3. Oldenburg LA191. 1 – 3 Charred emmer grains, *Triticum dicoccon*, dorsal and lateral views. 4 and 5 Charred seeds of crab apple, *Malus sylvestris*. 6 Charred nutshell fragments of hazel, *Corylus avellana*.

5.3 Oldenburg LA232

Site name	Oldenburg LA232 (settlement phase 1)
Longitude, latitude	54.291503, 10.853191
Federal state, country	Schleswig-Holstein, Germany
Excavation campaign	2016
Excavation type	Research excavation (SFB1266)
Responsible archaeologists	Jan Piet Brozio, Johannes Müller, Kiel
Archaeological context	Settlement
Analysed structures	Pits, postholes, cultural layers
Archaeological group	Funnel Beaker Culture
Archaeological periodisation	Middle Neolithic (MN I – MN II)
Absolute dating	3340 – 3016 cal BCE
Archaeobotanists	Dragana Filipović, Wiebke Kirleis, Kiel (SFB1266)
Archaeobotanical sampling strategy	Judgement sampling from archaeological contexts plus random sampling within 1 x 1 m quadrants of the grid
Analysed sediment volume	c. 380 l
Find concentrations (finds/l)	9 for the charred remains, c. 6 for the waterlogged remains
Number of carpological remains	3255 (charred), est. 2317 (waterlogged, extrapolated)

Fig. 5.3.1. Bird's eye view of the excavation, Oldenburg LA232 (photo: Jan Piet Brozio, UFG Kiel).

The Middle Neolithic settlement site (Table 8) of Oldenburg LA232 is situated on a sandy island in a wetland area in eastern Holsatia called Oldenburger Graben (Fig. 5.3.1), which, in Mesolithic and Neolithic times, was densely settled. Neighbouring sites include the Late Mesolithic and MN V site of Wangels LA505 (section 5.4)the EN II–MN IV site Oldenburg LA191 (section 5.2) and the MN II – IV site of Oldenburg LA77, Dannau (section 5.1). The Oldenburger Graben was cut off from the Baltic during the Neolithic; since then the settlements have been surrounded by marine to brackish waters. Excavations at Oldenburg LA232 concentrated on a transect extending from the uppermost part of the dune eastwards, towards the former shoreline (Brozio et al., 2018).

Insight into the plant economy and crop processing derives from the charred assemblages (Table 4). Emmer grain and glume bases represent the most abundant and frequent finds across the excavated contexts (Fig. 5.3.2). Einkorn and barley are rare finds and may represent accidental inclusions in stores of emmer. Threshing remains hint to storage of intact ears. Weed finds are small in absolute numbers and include *Chenopodium album, Bromus* sp., and other tiny Poaceae. Inerestingly, many charred remains were found in postholes, indicating ritual depostion of crop remains after house abandonment (Filipović et al., 2019). The waterlogged material reflects the composition of the natural vegetation along the former shoreline. The identified taxa suggest the presence of an ecosystem comprising of reservoirs of fresh, brackish, and/ or salt water, and the number of taxa demonstrates the high diversity of the vegetation. Such lush vegetation could have been grazed by herded animals or collected as fodder.

Fig. 5.3.2. Storage pit with mass find of emmer, Feature 3011, Oldenburg LA232 (photo: Jan Piet Brozio, UFG Kiel).

Literature:

Brozio, J.P., Filipović, D., Schmölcke, U., Kirleis, W., Müller, J., 2018. Mittel- bis Jungneolithische Siedlungshinterlassenschaften zwischen 3300 – 2600 v. Chr.—Der Fundplatz Oldenburg LA232 im Oldenburger Graben, Ostholstein. Prähistorische Zeitschrift 93(2), 185 – 224.

Filipović, D., Brozio, J.P., Kaczmarek, J., Müller, J., Kirleis, W., 2019 (forthcoming). Food transformed? Tapho-nomical investigation into a potentially symbolic role of crops at two Neolithic settlements in northern Germany, Prähistorische Zeitschrift.

Plate 1. Oldenburg LA232. 1 Charred emmer mass find, *Triticum dicoccon*. 2 Charred emmer mass find, zoomed in.

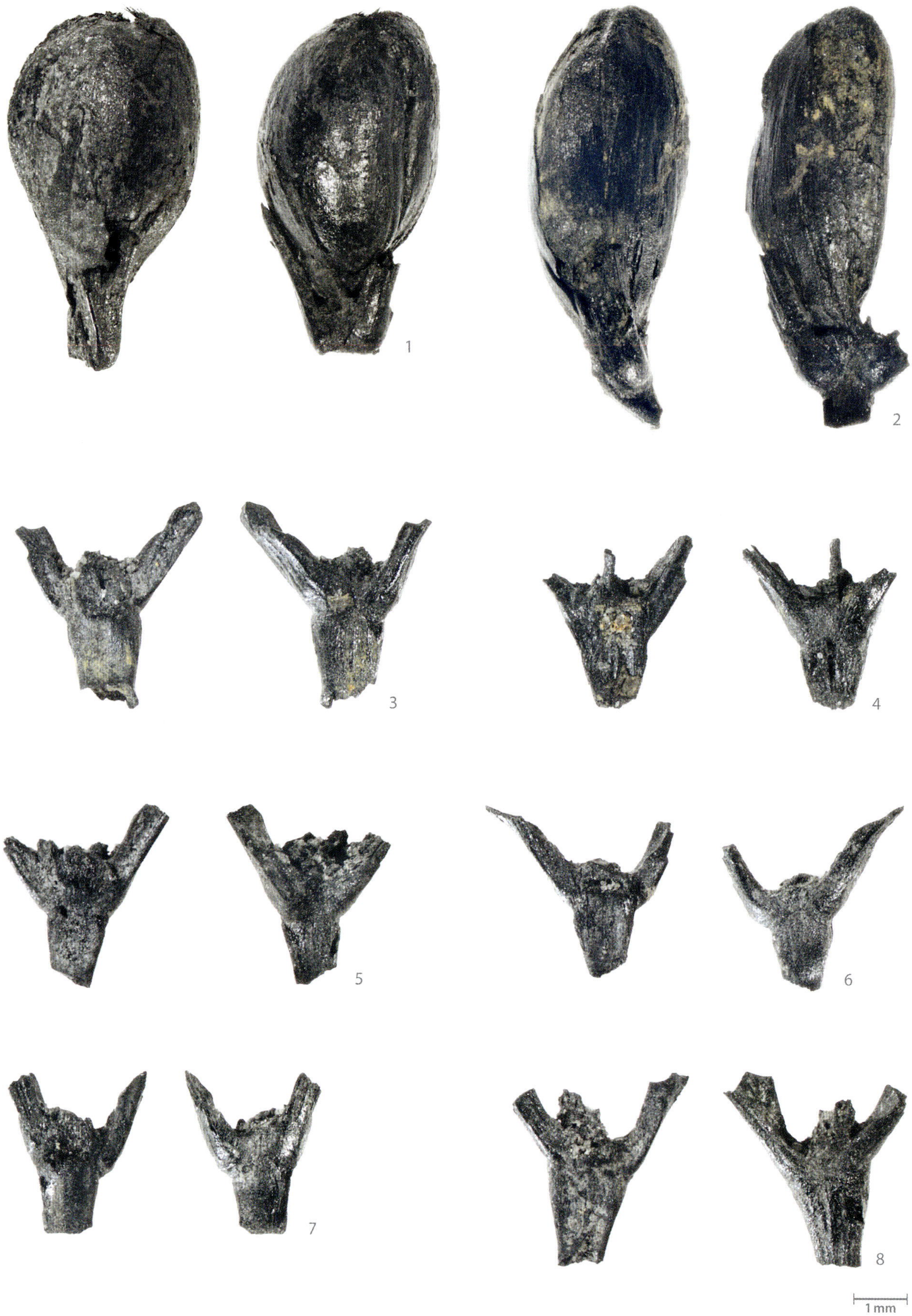

Plate 2. Oldenburg LA232. 1 and 2 Charred emmer grain, *Triticum dicoccon,* attached to spikelet fork, dorsal and lateral views. 3 – 8 Charred emmer spikelet forks with rachis segments, inner and outer views.

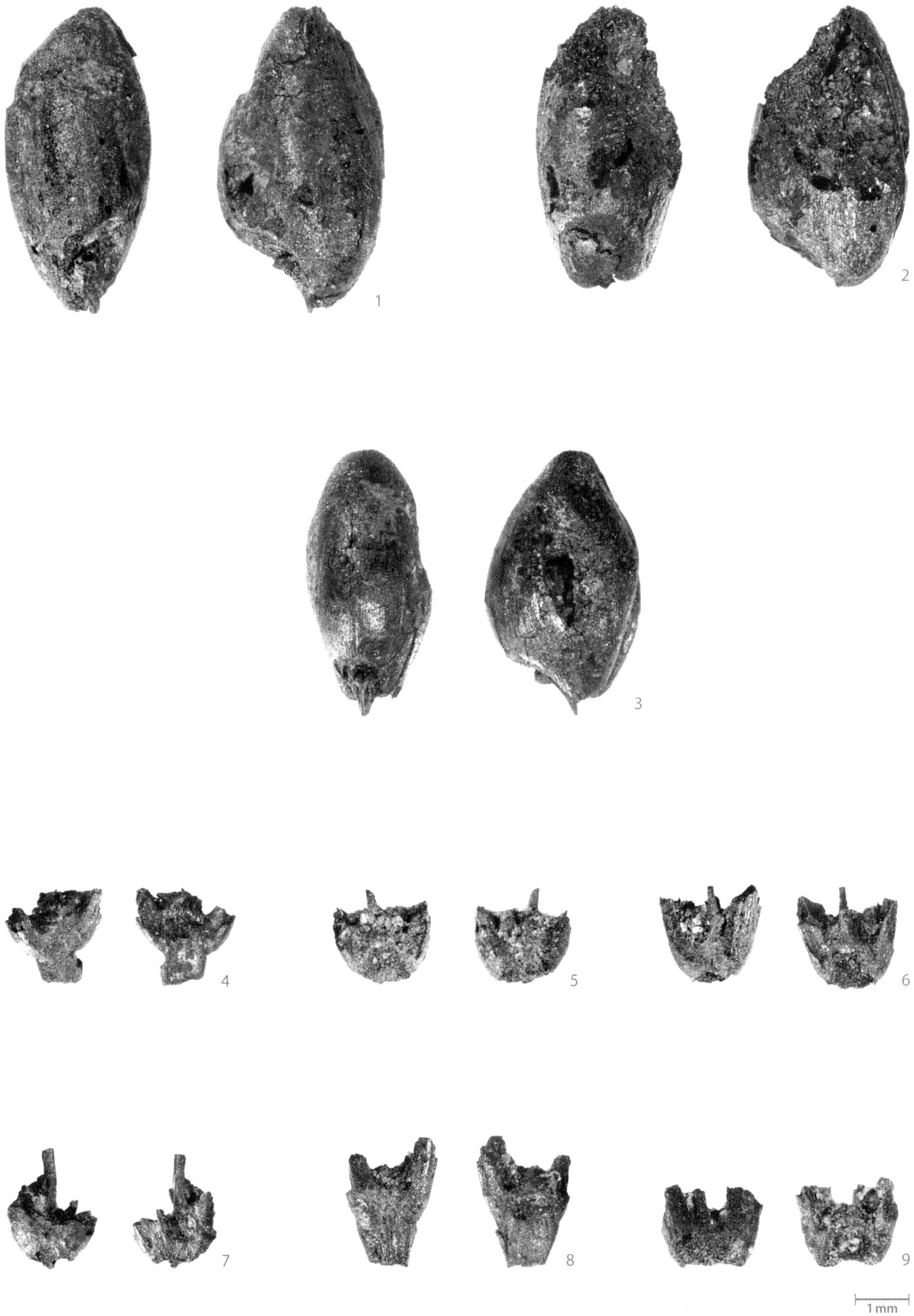

Plate 3. Oldenburg LA232. 1 – 3 Charred einkorn grains, *Triticum monococcum*, dorsal and lateral views. 4 – 9 Charred einkorn spikelet forks, inner and outer views.

Plate 4. Oldenburg LA232. 1–3 Charred grains of barley, *Hordeum vulgare*, dorsal and ventral views. 4 Charred rachis segments of barley. 5 and 6 Charred fragments of hazel nutshells, *Corylus avellana*, cross-section and surface. 7 Charred fruits of timothy grass, *Phleum pratense*. 8 Charred seeds of white goosefoot, *Chenopodium album*.

5.4 Wangels LA505

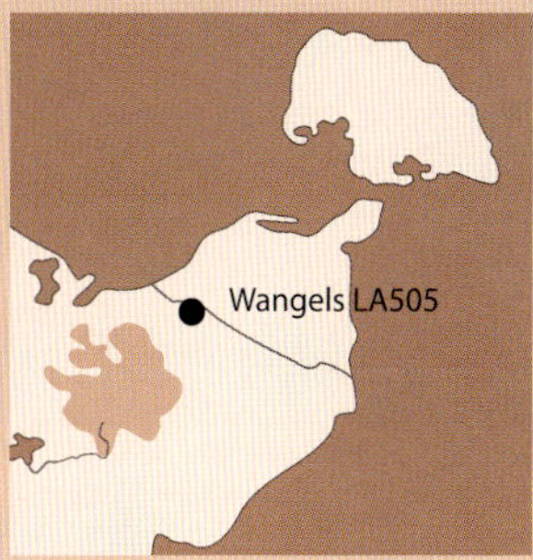

Site name	Wangels LA505
Longitude, latitude	54.291503, 10.853191
Federal state, country	Schleswig-Holstein, Germany
Excavation campaign	1996 – 1999
Excavation type	Research excavation
Responsible archaeologists	Sönke Hartz, Museum für Archäologie Schloss Gottorf; Ringo Klooß, Kiel
Archaeological context	Settlement
Analysed structures	Cultural layers, rubbish deposition in shallow water
Archaeological group	Funnel Beaker Culture
Archaeological periodisation	Middle Neolithic (MN V)
Absolute dating	2900 – 2800 cal BCE
Archaeobotanist	Helmut Kroll, Kiel
Archaeobotanical sampling strategy	Judgement sampling from archaeological contexts and from hand-picked material
Analysed sediment volume	c. 65 l
Find concentrations (finds/l)	147 (waterlogged); 2 (charred)
Number of carpological remains	9543 (waterlogged); 135 (charred) plus numerous threshing remains

The site of Wangels LA505 is located on a sandy island in the wetland area called Oldenburger Graben, which is situated in the younger morainic plain in eastern Holsatia (Fig. 5.1). In Mesolithic and Neolithic times, this area was densely settled. Neighbouring sites include the Middle Neolithic sites of Oldenburg LA77, Dannau; Oldenburg LA191; and Oldenburg LA232. The Oldenburger Graben was cut off the Baltic during the Neolithic, and since then the settlements have been surrounded by a marine to brackish water body, as is shown by palaeobotanical studies (Rickert, 2007). At the site of Wangels LA505, human activity dates to the transition from the late Mesolithic to the Early Neolithic and to the transition from the Middle to the Younger Neolithic period (Table 8; Klooß, 2008). The Early Neolithic artefacts comprise a broad variety of stone tools, including querns made of granite; rubbing stones made of sand stone; and diverse ceramics, including fragments of beakers of the Ertebølle pointed base type, as well as Funnel beakers and lamps. The dominance of mammal bones from domestic animals indicates the transition from fisher-hunter-gatherer to farming communities in this early stage. The waterlogged preservation conditions allowed for preservation of wooden artefacts, among which numerous eel spears made of hazel, Maloideae rods, and paddles made of ash wood, showing the affinity of the dwellers to the aquatic environment. Sharpened hazel rods 2 – 3 cm in diameter found sticking out of the ground in upright position are interpreted as the remains of a fish fence (Klooß, 2015). The late Middle to Younger Neolithic domestic site is characterized by finds of flint scrapers, transverse arrow heads, rubbing stones, pounders, and unornamented ceramic fragments of Store Valby type, including from ceramic plates with impressed edges. Artefacts of amber, wood, antler, and bones supplement the finds spectrum (Klooß, 2015, 35 – 55).

Archaeobotanical material was recovered from the cultural layers of the late Middle Neolithic period MN V only; soil samples from the earlier period did not contain any crop remains. The MN V assemblage is rich because plant remains were preserved under waterlogged conditions as subfossil remains and under dry conditions as charred remains (Kroll, 2001). Some of the material is archived at the Archäologisches Landesmuseum Schloss Gottorf, in Schleswig, of which exemplary material is documented here. The crop spectrum is dominated by emmer finds (Table 7). Barley, mainly the free-threshing variety, is the second most important cereal species. There were also single finds of free-threshing wheat. When documenting the material, two rachis segments could be allocated as free-threshing wheat with tetraploid genomic constitution *Triticum durum/turgidum*, which is known from Early Neolithic Funnel Beaker contexts (sections 3.4 and 7.2). The finds from Wangels LA505, although sparse in number, indicate that *Triticum durum/ turgidum* may have persisted in the north throughout the Early Neolithic and on to the late Middle Neolithic. The most important finds from Wangels LA505 are the seeds of opium poppy *Papaver somniferum*, which, in general, are rare finds because they easily burn to ash due to their high oil content. Their occurrence indicates the use of this oil plant in the Funnel Beaker period. Whether poppy was primarily used as an oil plant to supplement the diet or whether the dried latex was already being used to produce opium cannot be clarified from the seed finds. However, it cannot be excluded that the Neolithic dwellers were experienced in the use of drug plants; the sedative properties of the alkaloid-rich, opiate-containing latex of the opium poppy may well have been recognized and appreciated. In the area of the Oldenburger Graben, seeds of opium poppy have been found in a Funnel Beaker context at only two sites, Oldenburg LA77, Dannau (section 5.1), and Wangels LA505. Opium poppy is a crop plant of western European origin whose possible wild ancestor, *Papaver segetum*, grows in the western Mediterranean area. The cultivation of poppy spread from there towards central Europe in a second phase of the Linearbandkeramik Culture, after 5300 BCE, either as the domestic species or as the wild species, from where it was domesticated in multiple regions, possibly in central Germany and Belgium (Salavert, 2011). At Wangels LA505, the crops are supplemented by gathered plants, such as hazel, crab apple, hawthorn, wild strawberry, raspberry, and bramble. Finds of herbaceous plants can be subdivided into a weed assemblage and a group of wetland and aquatic plants indicative of the natural environment. Chenopods are most prominent in the weed assemblage; these are accompanied, among others, by knotweeds; grasses, such as *Phleum*, *Poa*, and *Bromus*; and broadleaf plantain.

Literature:

Klooß, R., 2008. Ein Fundplatz des Übergangs vom Mittelneolithikum A zum Mittelneolithikum B aus Wangels LA 505, Kreis Ostholstein. Ein Vorbericht, in: Dörfler, W., Müller, J. (Eds.), Umwelt, Wirtschaft, Siedlungen im dritten vorchristlichen Jahrtausend Mitteleuropas und Südskandinaviens. Offa-Bücher 84. Wachholtz, Neumünster, pp. 127 – 134.

Klooß, S., 2015. Mit Einbaum und Paddel zum Fischfang. Holzartefakte von endmesolithischen und frühneolithischen Küstensiedlungen an der südwestlichen Ostseeküste. Wachholtz, Kiel.

Kroll, H., 2001. Der Mohn, die Trichterbecherkultur und das südwestliche Ostseegebiet, in: Kelm, R. (Ed.), Zurück zur Steinzeitlandschaft. Archäobiologische und ökologische Forschung zur jungsteinzeitlichen Kulturlandschaft und ihrer Nutzung in Nordwestdeutschland 3. Albersdorfer Koll., Heide, pp. 70 – 76.

Rickert, B., 2007. Vegetationskundliche Interpretation botanischer Makroreste aus den ertebølle-zeitlichen und frühneolithischen Fundschichten des Siedlungsplatzes Wangels LA505. Schriften des Naturwissenschaftlichen Vereins Schleswig-Holstein 69, 15 – 28.

Salavert, A., 2011. Plant economy of the first farmers of central Belgium (Linearbandkeramik, 5200 – 5000 B.C.). Vegetation History and Archaeobotany 20(5), 321 – 332.

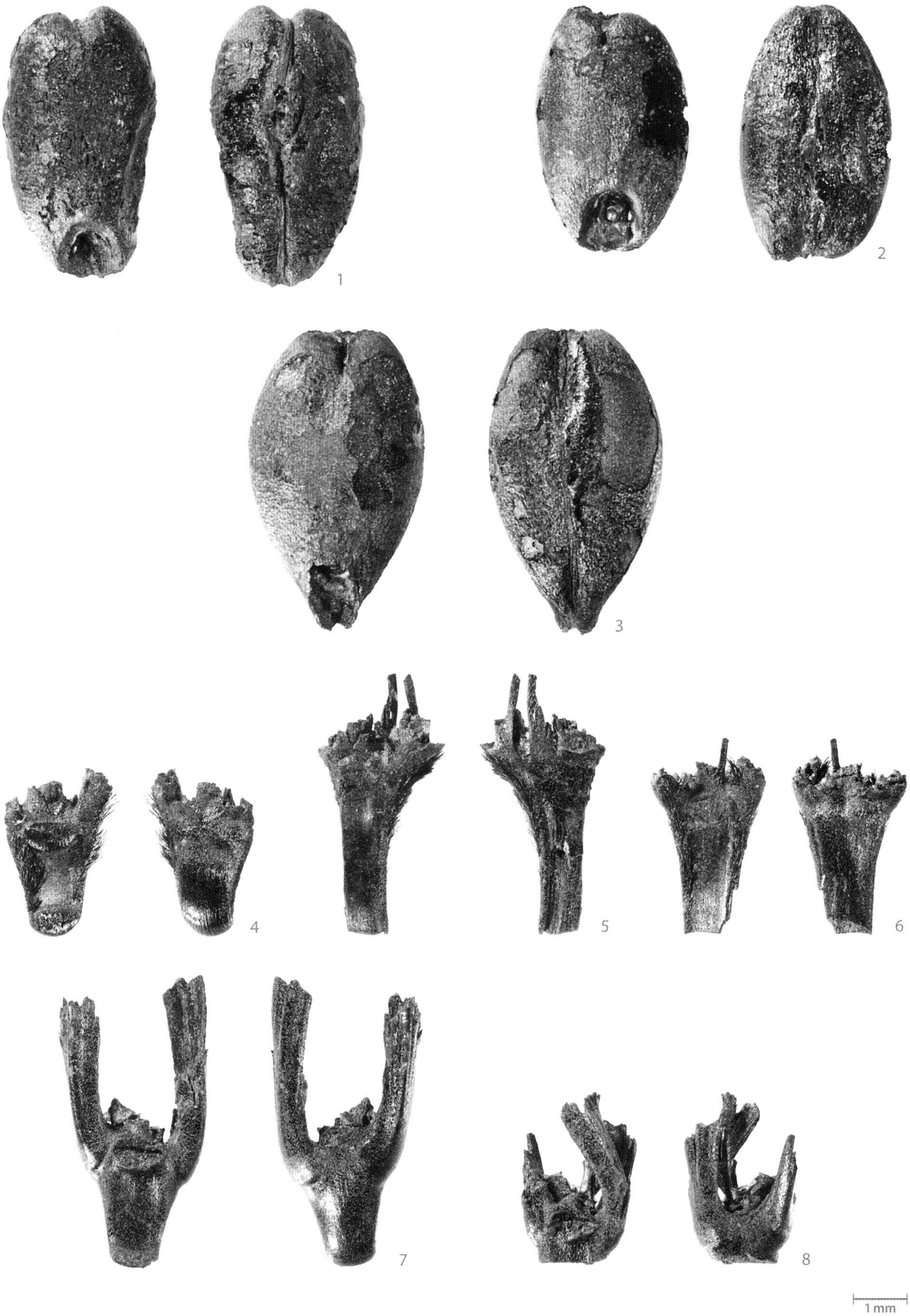

Plate 1. Wangels LA505. 1 – 3 Charred grains of free-threshing barley, *Hordeum vulgare nudum*, dorsal and ventral views. 4 – 6 Charred/subfossil rachis segments of barley, *Hordeum vulgare*, inner and outer views. 7 and 8 Charred/subfossil einkorn spikelet forks, *Triticum monococcum*, inner and outer views.

Plate 2. Wangels LA505. 1 Charred grain of free-threshing wheat, *Triticum aestivum/durum*, dorsal and ventral views. 2 Charred rachis segment of tetraploid free-threshing wheat, *Triticum durum/turgidum*. 3 Charred rachis segment of tetraploid free-threshing wheat, inner and outer views. 4 Subfossil seeds of opium poppy, *Papaver somniferum*, lateral and top views. 5 and 6 Charred rye brome grains, *Bromus secalinus*, dorsal and ventral views. 7 Charred seed of yellow iris, *Iris pseudacorus*.

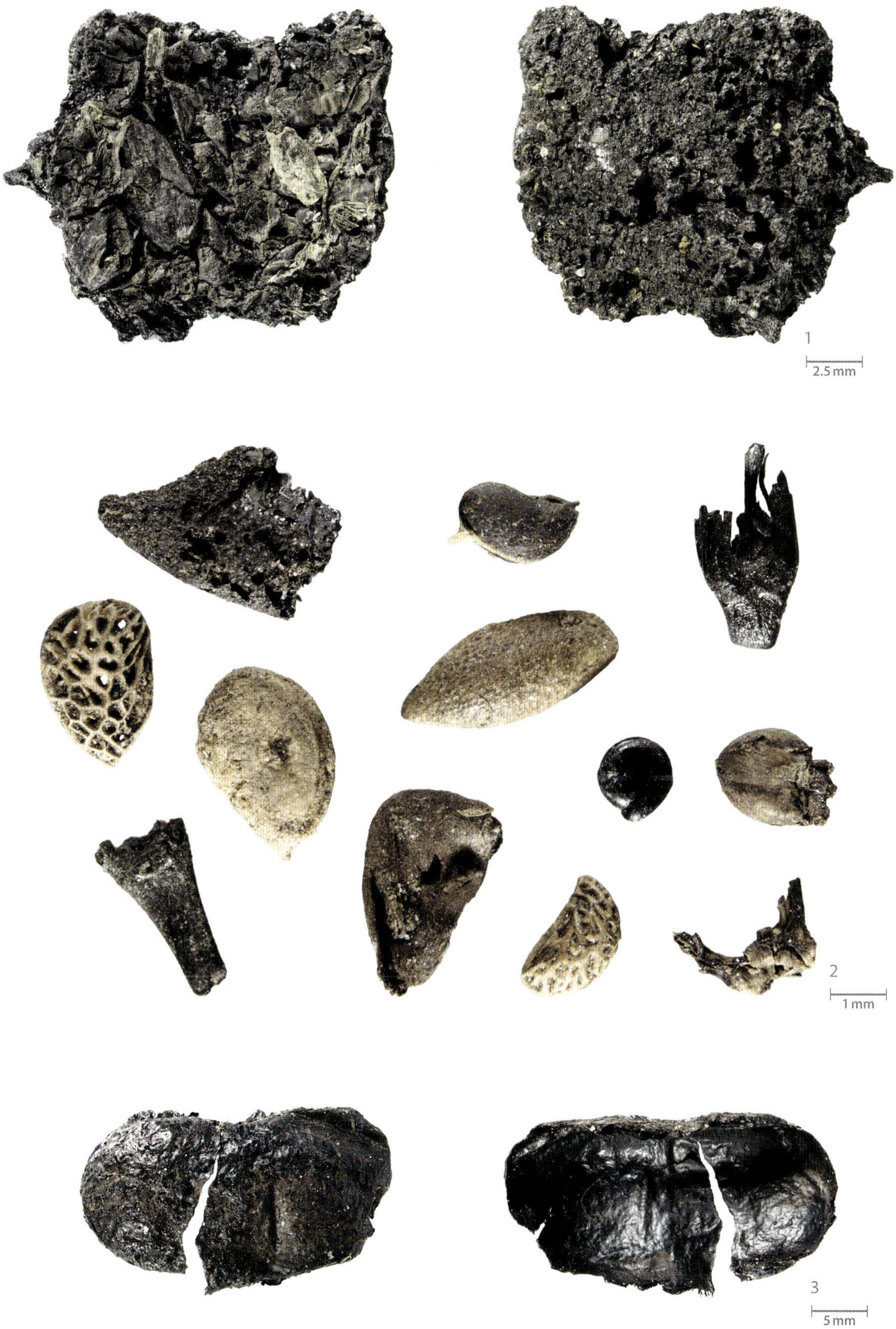

Plate 3. Wangels LA505. 1 In situ agglomeration. Top: subfossil remains, reverse view: charred. 2 Stray finds of subfossil material. Top: Einkorn, *Triticum monococcum,* beaked tasselweed, *Ruppia maritima,* einkorn spikelet fork with rachis. Centre: Bramble, *Rubus fruticosus,* pondweed, *Potamogeton* sp., holly-leaved naiad, *Najas marina,* white goosefoot, *Chenopodium album,* bud. Bottom: Barley, *Hordeum vulgare,* rachis segment, apple, *Malus* sp., raspberry, *Rubus idaeus,* emmer, *Triticum dicoccon,* spikelet fork. 3 Insect cocoon, outer and inner views.

5.5 Grube LA65

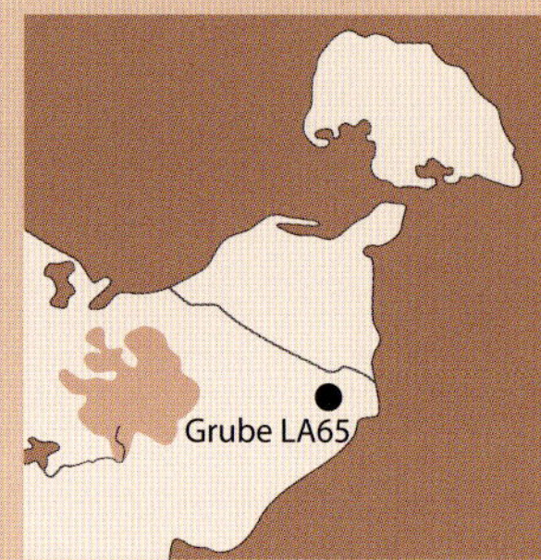

Site name	Grube LA65
Longitude, latitude	54.261892, 11.080546
Federal state, country	Schleswig-Holstein, Germany
Excavation campaign	2012
Excavation type	Rescue excavation
Responsible archaeologist	Ingo Lütjens, Archäologisches Landesamt Schleswig-Holstein
Archaeological context	Settlement
Analysed structures	Pit
Archaeological group	Funnel Beaker Culture
Archaeological periodisation	Early Neolithic (EN Ib – EN II)
Absolute dating	3650–3400 cal BCE
Archaeobotanists	Stefanie Klooß, Wiebke Kirleis, Kiel (SPP1400)
Archaeobotanical sampling strategy	Judgement sampling from archaeological contexts
Analysed sediment volume	10 l
Find concentrations (finds/l)	10
Number of carpological remains	104

Prior to dyke construction work in the eastern part of the wetland area Oldenburger Graben (Fig. 5.1), close to the village Rosenfelde, rescue excavations were carried out at the site Grube LA65. One Early Neolithic pit contained, in addition to flint artefacts and ceramic sherds, a mass find of charred seeds and fruits (Tables 4 and 8). Free-threshing barley dominates the cereal spectrum; emmer grains are represented by single finds. Some glume bases further indicate the occurrence of the hulled wheats emmer and einkorn. Nutshells of hazel and remains of crab apple supplement the food plants. Tiny fruits of Poaceae are the main component of the weed spectrum; vetches and goosefoot are represented by single finds.

Literature: –

Plate 1. Grube LA65. 1 Charred grain of free-threshing barley, *Hordeum vulgare nudum*, dorsal and ventral views. 2 Charred spikelet fork of einkorn, *Triticum monococcum*, inner and outer views. 3 Charred seed of goosefoot, *Chenopodium* sp. 4 Charred seed of vetch, *Vicia* sp. 5 Tiny caryopses of the grass family, Poaceae. 6 – 8 Tiny caryopses of the grass family, dorsal, lateral, and ventral views. 9 and 10 Annual meadow grass/timothy grass, *Poa annua/Phleum pratense*, dorsal, lateral, and ventral views.

5.6 Heringsdorf LA38, Süssau

Site name	Heringsdorf LA38, Süssau
Longitude, latitude	54.291503, 10.853191
Federal state, country	Schleswig-Holstein, Germany
Excavation campaign	1965–1967
Excavation type	Rescue excavation
Responsible archaeologist	Jürgen Hoika, Museum für Archäologie Schloss Gottorf
Archaeological context	Settlement
Analysed structures	Posts, pits
Archaeological group	Funnel Beaker Culture
Archaeological periodisation	Early Neolithic (EN Ib); Middle Neolithic (MN I/II)
Absolute dating	3650–3500 BCE; 3350–3050 BCE (according to new ^{14}C-dates)
Archaeobotanists	Elske Fischer, Wiebke Kirleis, Kiel (SPP 1400)
Archaeobotanical sampling strategy	Judgement sampling (on archived material)
Analysed sediment volume	24 l
Find concentrations (finds/l)	1 (0.8)
Number of carpological remains	19

The site of Heringsdorf LA38, Süssau, is located in the eastern part of the wetland area called Oldenburger Graben (Fig. 5.1). It is located in the lowlands, 200 m behind a beach ridge facing the bay of Lübeck. The date of the site was given as Middle Neolithic III/IV by Hoika (1972). New radiocarbon analyses applied within the research program SPP1400 suggest an Early Neolithic I b to Middle Neolithic I–II age for the site (Table 8). For EN I b, only nutshell fragments of hazel were found. The later Middle Neolithic shows a small archaeobotanical charred assemblage with poor preservation (Table 4). It comprises some cereal grains, among which the hulled wheat emmer, accompanied by single finds of black bindweed and white goosefoot.

Literature:

Hoika, J., 1972. Süssau, ein neolithischer Wohnplatz an der Ostsee. Archäologisches Korrespondenzblatt 2, 17–19.

Hoika, J., 1987. Das Mittelneolithikum zur Zeit der Trichterbecherkultur in Nordostholstein—Untersuchungen zu Archäologie und Landschaftsgeschichte. Offa-Bücher 61. Wachholtz, Neumünster.

Plate 1. Heringsdorf LA38, Süssau. 1 Charred fragment of hazel nutshell, *Corylus avellana*. 2 – 4 Charred emmer grains, *Triticum* cf. *dicoccon*, lateral view. 5 Charred fragment of black bindweed, *Fallopia convolvulus*. 6 Charred seeds of white goosefoot, *Chenopodium album*.

6 Mecklenburg-Vorpommern and Brandenburg: The upper moraine landscapes in the north-east

The geomorphology of the federal states of Mecklenburg-Vorpommern and northern Brandenburg is predominantly shaped by the Weichsel glaciation. The upper moraine landscape shows a highly dynamic relief, with smaller terminal moraines, glacial hollows that developed into lakes, dunes and sander areas, and glacial valleys with lowland peat bog development. The ground moraine is characterized by sandy loam and loamy sand soils with medium to good fertility (Benda, 1995; Lorenz et al., 2014). Western Mecklenburg is characterized by numerous freshwater bodies. This area was already a geographically favourable communications junction by Neolithic times due to the large, interconnected lakes around Schwerin; the six substantial rivers west of the Oder that connect with the Baltic Sea; and the three rivers running southwards connecting with the Elbe, which drains north-west into the North Sea. Farther south, the Havelland, in the federal state of Brandenburg, is a confluence of three major glacial river valleys, of the Baruther, the Berlin, and the Eberswalde. The meltwater flow led to deposition of fine sands and divided the former moraine plateau into small islands consisting of fragmented ground moraine plates, such as the Nauener Platte. Undulating ground moraines, small remains of terminal moraines, outwash plains, and dunes are located in the immediate vicinity of wetlands and the riverine landscape of the Havel, where peat growth started in the Late Pleistocene. This diversification is mirrored in a variety of soil types; hydromorphic soils prevail in the lowlands, while loamy sand or sandy loam with medium to good soil fertility are encountered on higher elevations of the ground moraine (Lutze 2014: 71).

Neolithization in the area of western Mecklenburg and Brandenburg was generally thought to have been initiated by the Funnel Beaker Culture (FBC) with the expression of the Baalberge groups in a late stage, in the fourth millennium BCE. However, stray finds of the Early Neolithic LBK and Middle Neolithic archaeological groups of the fifth millennium BCE are known from the regions of the Havelland as well as the Uckermark since the 20th century. The early expansion of a fully Neolithic way of life within Europe in the second half of the sixth millennium BCE is associated with the oldest central European farming group, the LBK. Traditionally, it is assumed that the area inhabited by LBK groups extended as a so-called Altsiedellandschaft in the loess belt, i.e. in the west up to the river Rhine and in the north up to the loess border north of the low mountain range. The northern European lowlands supposedly were the area of pervading Mesolithic people, who built their economy on hunting, fishing, and the collecting of wild plants as a preferred subsistence strategy. Crop cultivation and livestock breeding only became the basis for securing a livelihood in the lowlands about 1400 years later, with the onset of the FBC (e.g. Kirleis et al., 2012; Müller, 2011). However, recent excavations show that in particular the Havelland region was a contact zone between Neolithic farmers and Mesolithic hunter-fisher-gatherers from c. 5200 BCE (Wetzel, 2000). Such border regions are particularly important as engines of cultural exchange, as they are a direct contact zone between Mesolithic and Neolithic lifestyles. Outside the loess region, 23 settlements of the middle to the youngest phase of the LBK with house plans and pits can be found on raised ground moraine plates protruding from the glacial valley, but also on dunes and in the lowlands.

They are distributed like islands among sites of clear Mesolithic character (Ismail-Weber, 2017; Kirsch and Uhl, 1990). It is surprising to find early arable farmers in this landscape, with its rather poor soils on the dunes, and its sand dunes, larger basins, moors, and the river valley of the Havel. It is possible that the criteria for the choice of settlement sites here were the easy workability of the soils and the easy access to and availability of freshwater. Neolithic settlements are preferentially found on the fragmented ground moraine plates with loamy sand to sandy loam with medium to good soil fertility today. Thus, the early farmers who cultivated tiny arable plots did benefit from the small-scale diversity of the landscape (Lutze, 2014: 73; Stackebrandt and Lippstreu, 2010; Weiße, 2000). In addition to the Early Neolithic LBK settlements, further important sites in the region dating to the Middle Neolithic Rössen period include a field of pits of c. 100 m^2 at Wustermark 9, with finds of typical ceramics and the cranium of an aurochs, and an exceptional well feature at Dyrotz 37, close to Wustermark, consisting of a hollowed out oak trunk, apparently part of a well, with seven partly decorated vessels deposited at the bottom of the well (May, 2000; Wetzel, 2000). The close spatial interlocking of Mesolithic and Neolithic legacies in the federal state of Brandenburg suggests close contacts between groups with very different subsistence strategies. Whether there was a continuous Neolithic development from the LBK up to the Funnel Beaker period, i.e. from 5300–3800 BCE, is unknown.

The archaeobotanical investigations on Neolithic sites in north-eastern Germany compiled here allow for insight into the ritual societal sphere, including material from two Neolithic cemeteries in West Mecklenburg, with studies on the large Neolithic cemeteries of Friedrichsruhe (section 6.1) and Pastitzer Forst, Fpl. 53, on the island of Rügen (section 6.2). Moving southwards to the Havelland, in Brandenburg, the presentation of the plant assemblages from domestic sites with Groß Behnitz, Fdpl. 15, comprising both LBK and FBC features (section 6.3); and the three LBK sites of Wachow Niebede, Fdpl. 4 (section 6.4); Lietzow 2 and Lietzow 10 (section 6.5); allows tracing the commonalities and discrepancies in the crop spectrum of these different Neolithic farming communities. While there are similarities in the cereal and oil plant spectra, the pea is a crop plant that occurs in Brandenburg LBK context but not in Brandenburg FBC contexts.

Literature:

Benda, L., 1995. Das Quartär Deutschlands. Gebrüder Bornträger, Berlin, Stuttgart.

Ismail-Weber, M., 2017. ...100 km from the next settlement ... mobility of Linear Pottery groups in Brandenburg, north-eastern Germany, in: Scharl, S., Gehlen, B. (Eds.), Mobility in Prehistoric sedentary societies. Marie Leidorf, Rahden, pp. 75–117.

Kirleis, W., Klooß, S., Kroll, H., Müller, J., 2012. Crop growing and gathering in the northern German Neolithic: a review supplemented by new results. Vegetation History and Archaeobotany 21(3), 221–242.

Kirsch, R., Uhl, U., 1990. Neue frühneolithische Funde im Havelland. Ausgrabungen und Funde 35/2, 57–61.

Lorenz, S., Börner, A., Lampe, R., 2014. Ein Schnitt durch das Land. Geologische und landschaftsgeschichtliche Einblicke im Leitungsgraben, in: Landesamt für Kultur und Denkmalpflege Mecklenburg-Vorpommern (Ed.), Pipeline: Archäologie. Ausgrabungen auf den großen Ferngastrassen in Mecklenburg-Vorpommern. Landesamt für Kultur und Denkmalpflege Mecklenburg-Vorpommern, Schwerin, pp. 39–48.

Lutze, G.W., 2014. Naturräume und Landschaften in Brandenburg und Berlin. Gliederung, Genese und Nutzung. Bebra Wissenschaftsverlag, Berlin.

May, J., 2000. Bodendenkmalschutz und archäologische Forschungen entlang der Wublitzrinne, in: Potsdam, Brandenburg und das Havelland. Führer zu archäologischen Denkmälern in Deutschland 37. Theiss, Stuttgart, 196–1205.

Müller, J., 2011. Megaliths and Funnel Beakers: societies in change, 4100–2700 BC. 33rd Kroon-Voordracht. Stichting Nederlands Museum voor Anthropologie en Praehistorie, Amsterdam.

Stackebrandt, W., Lippstreu, L., 2010. Brandenburg. Landescharakter und geologischer Bau, in: Stackebrandt, W., Manhenke, V. (Eds.), Atlas zur Geologie von Brandenburg. Landesamt für Geowissenschaften und Rohstoffe Brandenburg, Cottbus, pp. 10–17.

Weiße, R., 2000. Glazialgeologisch-geomorphologischer Überblick, in: Potsdam, Brandenburg und das Havelland. Führer zu archäologischen Denkmälern in Deutschland 37. Theiss, Stuttgart, pp. 11–18.

Wetzel, G., 2000. Jungsteinzeit, in: Potsdam, Brandenburg und das Havelland. Führer zu archäologischen Denkmälern in Deutschland 37. Theiss, Stuttgart, pp. 39–52.

6.1 Friedrichsruhe, Fpl. 7 and Fpl. 8

Site name	Friedrichsruhe, Fpl. 7; Friedrichsruhe, Fpl. 8
Longitude, latitude	Fpl. 7: 53.511494, 11.701950; Fpl. 8: 53.511052, 11.703380
Federal state, country	Mecklenburg-Western Pomerania, Germany
Excavation campaign	Fpl. 7: 2012 – 2015; Fpl. 8: 2010 – 2013
Excavation type	Research excavation (SPP1400)
Responsible archaeologists	Fpl. 7: Friedrich Lüth, DAI Berlin, Timothy Darvill, Bournemouth University; Fpl. 8: Friedrich Lüth, DAI Berlin, Frank Nikulka, Daniela Hofmann, University Hamburg
Archaeological context	Burial mounds
Analysed structures	Soil mantle of the burial mounds, pits, palaeosol
Archaeological group	Funnel Beaker Culture; Single Grave Culture
Archaeological periodisation	Fpl. 7: EN II – MN II; Fpl. 8: EN Ia – b; YN1 – 2
Absolute dating	Fpl. 7: c. 3500 – 3000 cal BCE; Fpl. 8: c. 3950 – 3500 and 2850 – 2450 cal BCE
Archaeobotanists	Yasmin Dannath, Stefanie Klooß, Wiebke Kirleis, Kiel (SPP1400)
Archaeobotanical sampling strategy	Judgement sampling from archaeological contexts
Analysed sediment volume	Fpl. 7: c. 270 l; Fpl. 8: c. 194 l
Find concentrations (finds/liter)	Fpl. 7: 1 (0.1); Fpl. 8: 1 (0.1)
Number of carpological remains	Fpl. 7: 23; Fpl. 8: 27

The large megalithic cemetery of Friedrichsruhe is situated on a flat plateau formed of fluvio-glacial sands and gravels, in western Mecklenburg. Four excavation trenches were opened at Friedrichsruhe Fpl. 7 in 2012–2014, revealing a long barrow without grave chamber (grave 1) and a single grave mound (grave 2). Soil samples for archaeobotanical investigations were taken from the soil mantle of the burial mound and from palaeosols.

The charred archaeobotanical assemblage is small (Table 5). Typical for a Neolithic tomb is evidence for nutshell fragments of hazel. Linkages with the economic sphere are shown by a small number of cereal remains, such as grains of barley and grains and glume bases of emmer. Wild plant remains supplement the assemblage. One tiny grain of broomcorn millet and two charcoal fragments of hornbeam are very likely intrusions from more recent periods. This may also account for the find of a fragment of acorn, since acorns are generally common for the Late Neolithic to Early Bronze Age transition. However, the predominant mixed oak forests would have allowed for the use of acorns in Early and Middle Neolithic times. The charcoal assemblage indicates that in the Neolithic the mixed oak forest at Friedrichsruhe was composed of oak, pine, birch, hazel, and Pomoideae. Finds of the softwood species willow/poplar and alder indicate riverine woodland in the vicinity of the site.

The largest monument of the cemetery, Friedrichsruhe, Fpl. 8, is a megalithic tomb of about 65 m length that is oriented north-east to south-west. This monument was constructed in two main phases, with an early phase dating to the early fourth millennium BCE and a later phase dating to the early third millennium BCE (Table 8). Since the site was covered by woodland, the excavations encountered heavy disturbance by tree roots. Archaeological artefacts comprise ceramic fragments and an arrow head dating to the Early Funnel Beaker period. Several querns, partly withuse wear traces, were deposited in the grave mound.

The archaeobotanical assemblage here contains few cereal grains, mainly emmer and some glume bases of hulled wheat, plus one heavily eroded remain of legumes, the shape being reminiscent of pea (Table 5). The cultivars most possibly were deposited before the erection of the tomb. Radiocarbon dating performed in 2013 by the Leibniz laboratory at Kiel University dated one cereal grain to the Funnel Beaker period of the Early Neolithic Ia and further crops to around 3600 cal BCE, which is the period of the Early Neolithic Ib. These exceptionally early dates for the crops from Friedrichsruhe, Fpl. 8 have to be interpreted with utmost care since the cereal grains recovered here are pretty sandy and partly vitrified (Plate 1, Figs. 1 and 2, Plate 2, Fig. 1). Ancient carbon adhering to the sand may be responsible for a time offset and may have caused the dates to come back too old. The single EN Ib probable pea find is suspicious for a Funnel Beaker context; only the site of Rullstorf, Fdpl. 8, in Lower Saxony (section 2.2), shows evidence for potential peas and for seeds of vetches, but there they originate from windthrow pits and are thus not necessarily of Neolithic origin. Legumes are not considered to belong to the Funnel Beaker crop assemblage, as emmer and barley are the plants commonly encountered (Bakels, 2014). At Friedrichsruhe, Fpl. 8, remains of wild plants, such as nutshells of hazel, bramble seeds, and one bulb of false oat grass *Arrhenatherum elatius* var. *bulbosum*, belong to the Early Neolithic grave context and the Younger Neolithic structures. A single seed of hornbeam was identified as a more recent intrusion, and this was later confirmed by a radiocarbon date pointing to the Bronze Age (Table 8).

Literature:

Bakels, C., 2014. The first farmers of the Northwest European Plain: some remarks on their crops, crop cultivation and impact on the environment. Journal of Archaeological Science 51, 94–97.

Behrens, A., Darvill, T., Lüth, F., Nikulka, F., 2013. Kurze Fundberichte 2012. Neolithikum. Friedrichsruhe, Lkr. Ludwigslust-Parchim. Jahrbuch Bodendenkmalpflege in Mecklenburg-Vorpommern 60, 2012 (2013), 318.

Behrens, A., Darvill, T., Lüth, F., 2015. Kurze Fundberichte 2014. Neolithikum. Friedrichsruhe, Lkr. Ludwigslust-Parchim, Fpl. 7. Jahrbuch Bodendenkmalpflege in Mecklenburg-Vorpommern 62, 2014 (2015), 217.

Darvill, T., Lüth, F., 2019. Early monumentality and social differentiation: a case study in western Mecklenburg, in: Müller, J., Hinz, M., Wunderlich, M. (Eds.), Megaliths, societies, landscapes: early monumentality and social differentiation in Neolithic Europe. Proceedings of the international conference in Kiel, 16–20 June 2015. Habelt, Bonn, pp. 193–203.

Drafehn, A., 2011. Kurze Fundberichte 2010. Neolithikum. Friedrichsruhe, Lkr. Parchim, Fpl. 8. Jahrbuch Bodendenkmalpflege in Mecklenburg-Vorpommern 58, 2010 (2011), 280.

Drafehn, A., 2012. Kurze Fundberichte 2011. Neolithikum. Friedrichsruhe, Lkr. Parchim, Fpl. 8. Jahrbuch Bodendenkmalpflege in Mecklenburg-Vorpommern 59, 2011 (2012), 312–313.

Plate 1. Friedrichsruhe, Fpl. 7. 1–3 Charred barley grains, *Hordeum vulgare*, dorsal and ventral views. 4 Charred emmer grain, *Triticum dicoccon*, dorsal and lateral views. 5 Charred seed of bramble, *Rubus fruticosus*. 6 Charred fruit of the grass family, Poaceae.

Plate 2. Friedrichsruhe, Fpl. 8. 1 Charred emmer grain, *Triticum dicoccon*, dorsal and lateral views. 2 Charred nutshell fragment of hazel, *Corylus avellana*, outer view. 3 Charred seeds of raspberry, *Rubus idaeus*. 4 Charred fruit of climbing false buckwheat/copse-bindweed, *Fallopia dumetorum*.

6.2 Pastitzer Forst, Fpl. 53

Site name	Pastitzer Forst, Fpl. 53
Longitude, latitude	54.378106, 13.450545
Federal state, country	Mecklenburg-Western Pomerania, Germany
Excavation campaign	2015
Excavation type	Research excavation
Responsible archaeologist	Anja Behrens, UFG Kiel, DAI Berlin
Archaeological context	Long barrow
Analysed structures	Mound
Archaeological group	Funnel Beaker Culture
Archaeological periodisation	Early Neolithic (EN Ib)
Absolute dating	3700 cal BCE
Archaeobotanists	Yasmin Dannath, Wiebke Kirleis, Kiel
Archaeobotanical sampling strategy	Judgement sampling from archaeological contexts
Analysed sediment volume	c. 100 l
Find concentrations (finds/l)	1 (0.3)
Number of carpological remains	28

Ritual landscapes of Neolithic origin characterize the southern and the eastern parts of the island of Rügen, in the Baltic Sea. Numerous monumental burials were erected by Funnel Beaker groups over the course of the fourth millennium BCE. Close to Putbus, an undisturbed group of about 23 burials comprising long barrows as well as round barrows was identified on Lidar scans in 2015 (Behrens, 2016). The burials line up along a sunken road over an area of 1.5 km^2. Excavations took place on one long barrow 37 m in length and 10 m in width (Fig. 6.2.1). Its short side was enclosed by erratic blocks. In contrast, erratic boulders and traces thereof are missing along its long side. Instead, partly stacked lines of medium-size stones were detected. Some archaeological artefacts dating to the Funnel Beaker period, such as flint tools and ceramic sherds, were found in the soil mantle of the grave mound. During the Iron Age, three well-preserved urns were deposited in the grave mound, and these are interpreted as secondary burials. Two radiocarbon dates on nutshells of hazel show that the gathered plants are contemporaneous with the Neolithic monument (Table 8). A radiocarbon date on a barley grain of c. 1000–800 cal BCE and the occurrence of a broomcorn millet grain, *Panicum miliaceum*, shows that this site is affected by intrusion of Bronze Age plant remains. The Iron Age urn depositions may account for even younger intrusions.

The archaeobotanical assemblage is pretty small (Table 5). In addition to the Neolithic nutshell fragments of hazel, single grains of barley and some weed remains were recovered. Since the plant assemblages of other Funnel Beaker tombs (e.g. Albersdorf LA5, Brutkamp, section 3.3) often are dominated by gathered plants, indicating that specific preferences may account for deposition

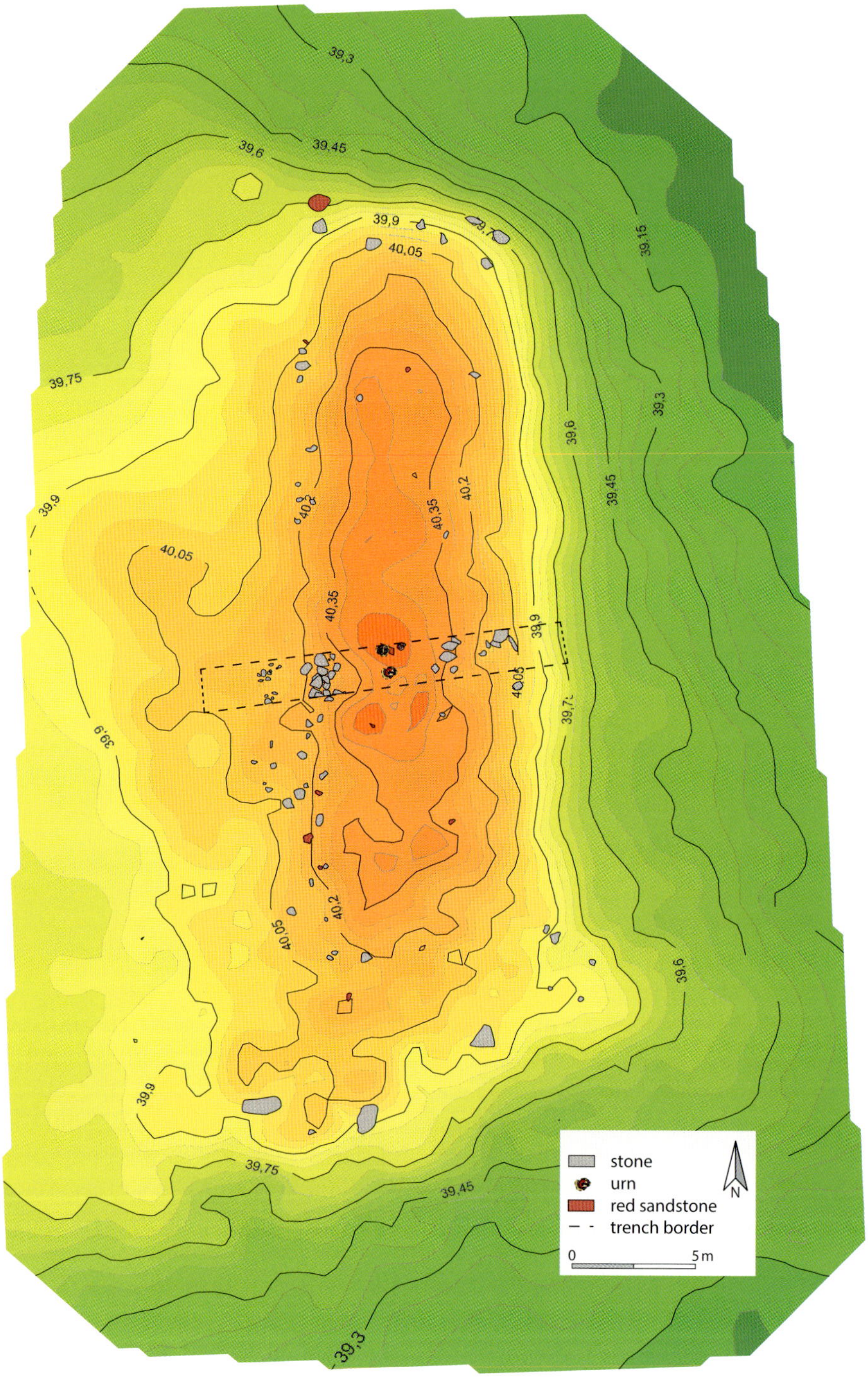

Fig. 6.2.1. Topography of the excavated Neolithic long barrow, with the excavation trench marked with a dashed line, Pastitzer Forst, Fpl. 53 (after Behrens, 2016: Fig. 4).

of plant remains of wild species in ritual contexts (Kirleis and Klooß, 2014), it remains unclear but probable that the cereal grains and weed remains detected here all are younger intrusions that date to the Bronze and Iron ages.

Literature:

Behrens, A., 2016. Rügen, Deutschland. Archäologische Untersuchungen an Großsteingräbern im Pastitzer Forst, in: E-Forschungsberichte des Deutschen Archäologischen Instituts, Faszikel 2. Deutsches Archaeologisches Institut, Berlin, pp. 24 – 27.

Kirleis, W., Klooß, S., 2014. More than simply fallback food? Social context of plant use in the northern German Neolithic, in: Chevalier, A., Marinova, E., Peña-Chocarro, L. (Eds.), Plants and people: choices and diversity through time. Earth Series 1. Oxbow Books, Oxford, pp. 326 – 335.

Plate 1. Pastitzer Forst, Fpl. 53. 1 Charred nutshell fragments of hazel, *Corylus avellana*. 2 Charred brome grain, *Bromus* sp., ventral and dorsal views. 3 Charred fragment of curlytop knotweed seed, *Persicaria lapathifolia*. 4 Charred lady's thumb seed, *Persicaria maculosa*. 5 Charred seed of black bindweed, *Fallopia convolvulus*. 6 Charred seed of the pink family, Caryophyllaceae, inner and outer views.

6.3 Groß Behnitz, Fdpl. 15

Site name	Groß Behnitz, Fdpl. 15
Longitude, latitude	52.578986, 12.769345
Federal state, country	Brandenburg, Germany
Excavation campaign	1997
Excavation type	Rescue excavation
Responsible archaeologist	Marion Dijkman-Dulkes, BLDAM, Wünsdorf
Archaeological context	Settlement
Analysed structures	Pits, postholes
Archaeological group	Linearbandkeramik; Funnel Beaker Culture
Archaeological periodisation	Early Neolithic, Late Neolithic (central European chronology)
Absolute dating	5200–4900 cal BCE; 3800–3600/3500 cal BCE
Archaeobotanists	Yasmin Dannath, Wiebke Kirleis, Kiel
Archaeobotanical sampling strategy	Judgement sampling from archaeological contexts
Analysed sediment volume	LBK: 55 l; FBC: 3 l
Find concentrations (finds/l)	LBK: 1 (0.3); FBC: 20
Number of carpological remains	LBK: 18; FBC: 61 and 1537 threshing remains

Groß Behnitz, Fdpl. 15, is situated on the ground moraine island of the Nauener Platte in Brandenburg's Havelland. In this area outside the loess belt, which is traditionally perceived as a region inhabited by Mesolithic hunter-fisher-gatherers, numerous Neolithic domestic sites have been identified, belonging to the Early Neolithic LBK groups, the Middle Neolithic Michelsberg groups, and the younger Neolithic FBC groups, such as the Tiefstich and Salzmünde groups. Thus, the area was a contact zone that was recurrently revisited by Mesolithic groups as well as early farmers, from the advanced LBK throughout the Neolithic. Especially noteworthy is the identification of nine postholes enclosing an area of 40 m², forming part of a LBK house structure with adjacent long pits at Groß Behnitz, Fdpl. 15. LBK pottery fragments and burnt daub comprise the artefact spectrum. Further structures at the site include features of the FBC, among others an agglomeration of five pits. Ceramic sherds, an arrow head, and burnt (animal?) bones belong to the younger period.

The small assemblage of LBK archaeobotanical finds originates from one feature only, which is the long pit, Feature 56, adjacent to the LBK house structure (Table 5). Among the 16 cereal grains two were determinable as emmer. One glume base each of emmer and einkorn adds to the crop remains. Single seeds of black bindweed and common knotweed comprise the weed remains. Although representing a small assemblage from soil samples with the low find concentrations that are typical for the Neolithic, these plant remains provide a first hint of local cultivation and processing of cereals in the upper moraine landscape of the Havelland in the late 2nd half of the sixth millennium BCE. This evidence clearly indicates the need for research excavations on LBK sites in this region with integration of intensive archaeobotanical studies to deliver a sound database to further confirm early farming activity by the LBK groups.

The FBC archaeobotanical finds at Groß Behnitz, Fdpl. 15, originate from a 3 l soil sample from Feature 52, which is the above-mentioned agglomeration of five pits (Table 5). Here the find concentrations, with about 20 carpological remains per litre, are pretty high. Similar to the LBK assemblage, emmer is the main cereal (Fig. 6.3.1), as indicated by grains and threshing remains. Some caryopses of *Bromus arvensis* and single seeds of common knotweed and the carnation family comprise the weed spectrum. The more than 1500 glume bases of hulled wheat, mainly of emmer, with some of einkorn, dominate the material in sheer numbers. Overall, the FBC archaeobotanical remains indicate an activity area where dehusking, being the last step of cereal processing, took place, whereafter the threshing remains were disposed of. Most probably, the hulled wheats were stored in the form of spikelets and dehusking was carried out prior to daily food preparation.

Fig. 6.3.1. Emmer grown in an experimental plot at Albersdorf. Ears with the right-angled turned uppermost spikelet (photo: Susanne Beyer, UFG Kiel, July 2017).

Literature:

Dijkman-Dulkes, M., 1998. Archäologische Untersuchung eines neolithischen Siedlungsplatzes im Bereich der ICE-Trasse in Groß-Behnitz, Kreis Nauen. Land-Bericht 6.

Jahns, S., Alsleben, A., Bittmann, F., Brande, A., Christiansen, J., Dannath, Y., Effenberger, H., Giesecke, T., Jäger, K.-D., Kirleis, W., Klooß, S., Kloss, K., Kroll, H., Lange, E., Medović, A., Neef, R., Stika, H.-P., Sudhaus, D., Wiethold, J., Wolters, S., 2018. Zur Geschichte der nacheiszeitlichen Umwelt und der Kulturpflanzen im Land Brandenburg, in: Beiträge zur Archäozoologie und Prähistorischen Anthropologie XI. Beier & Beran, Langenweißbach, pp. 9–35.

May, J., 2001. Ergebnisse bodendenkmalpflegerischer Aktivitäten entlang der Schnellbahnverbindung Hannnover-Berlin im Land Brandenburg. Beiträge zur Denkmalpflege und Forschung am Denkmal: Bodendenkmalpflege und Archäologisches Landesmuseum, 511–515.

Plate 1. Groß Behnitz, Fdpl. 15. 1 Charred wheat grain and wheat grain fragments, *Triticum* sp. 2 Charred glume bases of einkorn, *Triticum monococcum*.

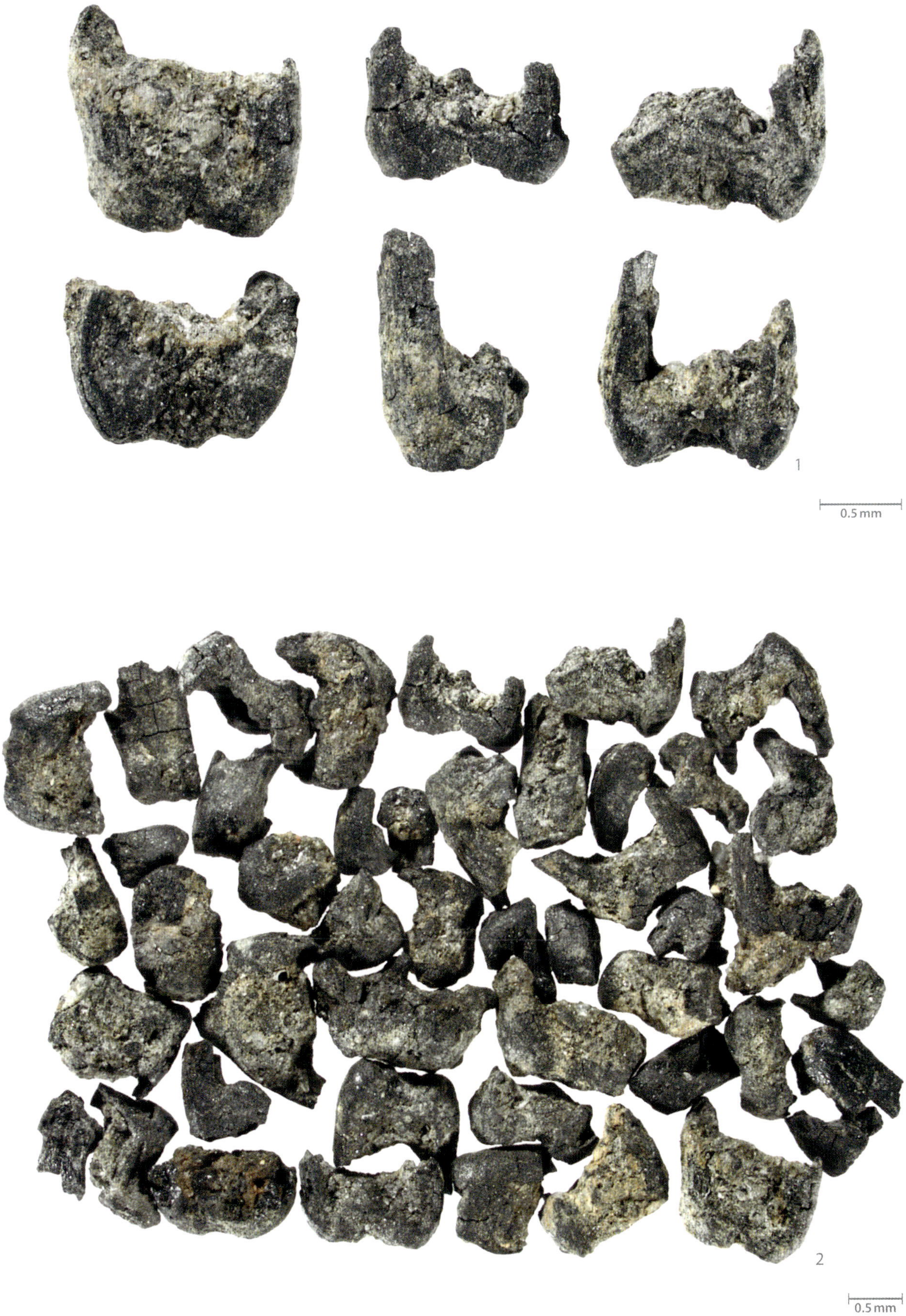

Plate 2. Groß Behnitz, Fdpl.15. 1 Charred glume bases of einkorn, *Triticum monococcum*. 2 Charred glume bases of einkorn/emmer, *Triticum monococcum/dicoccon*.

6.4 Wachow Niebede, Fdpl. 4

Site name	Wachow Niebede, Fdpl. 4
Longitude, latitude	52.540738, 12.787861
Federal state, country	Brandenburg, Germany
Excavation campaign	1996
Excavation type	Rescue excavation
Responsible archaeologist	Wolfgang Bünnig, Nauen
Archaeological context	Settlement
Analysed structures	Pit
Archaeological group	Linearbandkeramik
Archaeological periodisation	Early Neolithic
Absolute dating	5200–4900 cal BCE
Archaeobotanists	Yasmin Dannath, Wiebke Kirleis, Kiel
Archaeobotanical sampling strategy	Judgement sampling from archaeological contexts
Analysed sediment volume	22 l
Find concentrations (finds/l)	3
Number of carpological remains	72

Niebede is located in Brandenburg's Havelland region, between the Havelländisches Luch to the north and the river Havel to the south. The Linearbandkeramik site of Wachow Niebede, Fdpl. 4, is situated on a ground moraine plateau with soils of silty, fine sand at a presumably silted-up section of a lake called the Riewendsee. Prior to construction work, rescue excavations revealed early Neolithic cultural layers and pits with domestic waste (Bünnig, 1998). Archaeological artefacts comprise LBK ceramic fragments, flint tools, burnt daub, partly burned animal bones, a flat axe made of grey schist, a grinding stone made of sandstone, and bone awls. The LBK ceramic inventory consists of round-bottomed jars, bowls, and flask with a narrow bottleneck and horizontal handles. The material represents typical household goods, since it comprises fine ware as well as coarse ware.

The archaeobotanical assemblage contains typical LBK crop and weed remains. The hulled wheat emmer is the dominant cereal, evidenced mainly by threshing remains (Table 5). Barley supplements the cereal spectrum. While the single rye grain may be considered as belonging to the Neolithic weed remains, the single broomcorn millet grain presumably is a younger intrusion (Filipović et al., 2018). Pea represents the legumes, and several linseeds indicate the use of this oil plant at this LBK domestic site. The weed spectrum comprises white goosefoot, black bindweed, curly dock, barnyard grass, and brome, possibly rye brome, the latter being a species with the potential to be grown as crop. However, it was never established as a domestic plant (Bakels, 2014). The tiny charred awns of feather grass indicate that open areas with steppe vegetation most probably existed in the Neolithic woodlands of the Havelland region. The feather grass, which can become 70 cm tall, grows as a component of the natural vegetation in dry or lean grasslands. It is the character species of the continental steppe and dry grassland associations (Festucion and Xerobromion). Aesthetically, it impresses, particularly by its white feathered awns (Fig. 6.4.1). The long awns are spread easily by e.g. adhering to animal fur. Often awns mass together and form balls that are driven over open surfaces by the wind. In the Neolithic, feather grass obviously occurred in the surroundings of Niebede in dry places in open, sun-exposed, special locations. Most possibly, the concurrence of a continental micro-climate; nutrient-poor, sandy soils; and grazing by LBK domestic stock contributed to maintaining small-scale meadow steppe plots with feather grass and various flowering herbs. Ancient feather grass awns from Early Neolithic LBK and Middle Neolithic Stichbandkeramik group contexts are known from central Germany (Dannath et al., 2019; Kreuz, 2012). A targeted probable use of feather grass is indicated by finds of hundreds of awns and fruit stands laid down in neat rows in a Middle Neolithic pit at Schöningen, in central Germany. Its rolled leaves may have been used for the production of ropes and braids and as yarn for weaving. The lower end of the husk fruit, which is needle-sharp and hard, may even have been used as a tool or fine arrowhead for hunting small birds (Körber-Grohne, 1987).

Fig. 6.4.1. *Stipa capillata* grassland, Chechelnyk district, Ukraine (photo: Thomas Becker, Trier).

Literature:

Bakels, C., 2014. The first farmers of the Northwest European Plain: some remarks on their crops, crop cultivation and impact on the environment. Journal of Archaeological Science 51, 94 – 97.

Bünnig, W., 1998. Frühneolithische Besiedlung im Havelland. Eine Siedlungsgrube mit Bandkeramik in Niebede, Ot. von Wachow, Landkreis Havelland. Archäologie in Berlin und Brandenburg 1997. Theiss, Stuttgart, pp. 33 – 34.

Dannath, Y., Heske, I., Kirleis, W., 2019. Ein weiter Blick in die Landschaft. Entdeckungen an einem Standort der Stichbandkeramik. Archäologie in Niedersachsen 22, 50 – 54.

Filipović, D., Meadows, J., Wiethold, J., Jahns, S., Bittmann, F., Kirleis, W., 2018. Before and after: millet cultivation and the transformation of prehistoric crop production in northern Germany. Antiquity 92 (365), E3.

Körber-Grohne, U., 1987. Federgras-Grannen (*Stipa pennata* L. s. str.) als Vorrat in einer mittelneolithischen Grube in Schöningen, Landkreis Helmstedt. Archäologisches Korrespondenzblatt 17, 463 – 472.

Kreuz, A., 2012. Die Vertreibung aus dem Paradies? Archäobiologische Ergebnisse zum Frühneolithikum im westlichen Mitteleuropa. Berichte der Römisch-Germanischen Kommission 91. Zabern, Mainz, 23 – 196.

Plate 1. Wachow Niebede, Fdpl. 4. 1 Charred linseed, *Linum usitatissimum*. 2 Charred pea, *Pisum sativum*. 3 Charred spikelet fork of emmer, *Triticum dicoccon*. 4 Charred brome grains, *Bromus* sp.

Plate 2. Wachow Niebede, Fdpl. 4. 1 Charred seeds of white goosefoot, *Chenopodium album*. 2 Charred seed of curly dock, *Rumex crispus*. 3 Charred awn fragments of feather grass, *Stipa* sp. 4 Charred broomcorn millet grain, *Panicum miliaceum*, younger intrusion. 5 Charred rye grain, *Secale cereale*, younger intrusion or weed, dorsal and ventral views.

6.5 Lietzow 2 and Lietzow 10

Site names	Lietzow 2 and Lietzow 10
Longitude, latitude	Lietzow 2: 52.613660, 12.817259; Lietzow 10: 52.614676, 12.82557
Federal state, country	Brandenburg, Germany
Excavation campaigns	Lietzow 2: 2007, 2014; Lietzow 10: 2012
Excavation type	Rescue excavation
Responsible archaeologists	Ursula Uhl, Jonas Beran, BLDAM Wünsdorf; Andrea Weishaupt-Hahn, WHP-Archäologiebüro GbR, Berlin
Archaeological context	Settlement
Analysed structures	Pits, kiln
Archaeological group	Linearbandkeramik
Archaeological periodisation	Early Neolithic
Absolute dating	c. 5100–4900 cal BCE
Archaeobotanists	Yasmin Dannath, Kiel; Susanne Jahns, BLDAM Wünsdorf; Reinder Neef, DAI Berlin
Archaeobotanical sampling strategy	Judgement sampling from archaeological contexts
Analysed sediment volume	Lietzow 2: 129 l Lietzow 10: 130 l
Find concentration (finds/l)	Lietzow 2: 10 Lietzow 10: 35
Number of carpological remains	Lietzow 2: 1254 Lietzow 10: 4578

On the moraine plate of the Nauener Platte, in the Havelland region, rescue excavations uncovered two LBK domestic sites, Lietzow 2 and Lietzow 10. The ceramic typology indicates a younger Early Neolithic Age (5100 BCE onwards) for both and shows high variability, with influences from the northern foreland of the Harz mountain area, Saxonia, Lower Saxony, Bohemia, and the Moselle valley. The site of Lietzow 2 is located on the Mühlenberg, an area elevated about 50 m msl, with a steep slope towards the peaty wetland area known as the Havelländisches Luch, about 500 m west of the village of Lietzow. Stagnant ice holes in the lowlands close to Lietzow may have served as a freshwater reservoir for the LBK farmers. Lietzow 2 is a domestic site with 20 flat, heavily eroded LBK pits of different sizes, with loamy, partly humic fill. The archaeological artefacts comprise flint and stone tools, including fragments of stone adzes and a quern, and LBK ceramic fragments. Since the pits were mainly located in marl layers, bone preservation within them was good. Well-preserved, partly charred bones of cattle, pig, sheep, red deer, roe deer, terrapin, goose, and fish were recovered. One pit was laid out with a pavement made of boulders integrated with large fragments of a ceramic vessel and containing numerous fragments of bone and charcoal. This pit obviously was not used for waste disposal, since flint and ceramic fragments were absent.

It is suggested that this feature may have been a cooking pit, although there are no traces of localized burning in the sediment (Uhl 2008). The site of Lietzow 10, located at the western border of the village of Lietzow, comprises 22 features, mainly pits, of LBK archaeological context. Sherds of one vessel show the typical decoration of the Šarka region in Bohemia, of a pitched ceramic with triangular decoration made with birch bark or bone. Most of the ceramics are of local origin, but they imitate decoration from the core region of the LBK and thus indicate far-reaching social networks. Imported wares are rare; only one Limburgian vessel has a colour and texture that deviates completely from the local ceramic production. The silex industry at Lietzow 10 combines Neolithic and Mesolithic elements. Animal bones seem to show a high relevance of game, with indications for robust cattle bones, red deer, horse, bear, fox, rodents, diverse birds, and terrapin. Fish remains and animal tooth pendants supplement the archaeozoological remains. Overall, the archaeological artefacts from Lietzow 10 indicate a coalescence of farming and foraging traditions what is in good accordance with a site located in a Mesolithic–Neolithic borderland with intensive cross-cultural contacts. An extraordinary feature at

Lietzow 10 is a pit filled with clearly separated, alternating layers of light and dark fill, which run concentrically, like the layers of an onion (Fig. 6.5.1). The feature is interpreted as the remains of an earthen kiln. The dark layers contained the larger part of the finds, comprising ceramics, flint artefacts, charred plant remains, and partly burnt animal bones. According to the excavator, this suggests that the black layers originate from the use of the kiln, which were partly cleared out and then covered with clay for the next use. This could mean that each dark layer is the remnant of a single event (Hahn-Weishaupt 2014).

The archaeobotanical assemblage from Lietzow 2 is dominated by weed species, with more than a thousand seeds and fruits (Table 5). Among the 47 cereal caryopses, emmer is the only identifiable species. Glume bases of both emmer and einkorn further prove the presence of hulled wheat species. One grain shows the sunken embryo typical for a free-threshing wheat species (Plate 1, Fig. 2). Single peas supplement the crop spectrum. One fragment of a nutshell of hazel represents the gathered plants. The more than 1000 seeds of

Fig. 6.5.1. Stratified feature, Lietzow 10 (photo: Hahn-Weishaupt, Berlin).

white goosefoot and 145 fruits of the climbing black bindweed may have been a welcome addition to the daily nutrition or may simply represent disposal of contaminants of the cereal crop. If these species grew together with the cereals on arable plots, the weeds' growth height would indicate a cereal harvest at the level of the ears, since low-growth weed species are missing, except for two seeds of common sheep sorrel. The archaeobotanical assemblage from Lietzow 10 (Jahns et al., 2018; Kirleis et al. in prep.), with more than 4500 finds, originates from the stratified feature. Similar to Lietzow 2, the main cereal is the hulled wheat emmer, here evidenced by grains and glume bases, and supplemented by few einkorn grains. Pea and linseed complete the crop spectrum. A total of 17 different herbs are proven, represented by a total of 1500 seeds and fruits. More than 200 tiny seeds of varying size classes with mitaform seed shape and a divergent radicular lobe belonging to the legume family were identified as cf. *Melilotus* (Plate 4, Fig. 3 – 5). This genus is a rare occurrence for the Neolithic in Germany. *Melilotus alba* was found e.g. at the LBK site of Vaihingen/Enz (Rösch 1999). Many of the weed species may have been plants that were intentionally harvested, like the black bindweed or the brome species. Among the brome species, the slender caryopses of barren brome, with lengths about 8 mm, were determined (Plate 4, Figs. 1 and 2). Charred caryopses of barren brome from an assemblage of waterlogged plant remains from the Mesolithic campsite of Friesack 4, some 20 km northwest of Lietzow, were interpreted as potential plant food (Wolters 2006).

Literature:

Hahn-Weishaupt, A., 2014. Bauern, Jäger oder beides. Siedlung der Linienbandkeramik in Lietzow, Lkr. Havelland, in: Archäologische Gesellschaft in Berlin und Brandenburg e. V. in Zusammenarbeit mit dem Brandenburgischen Landesmuseum für Ur- und Frühgeschichte und dem Landesdenkmalamt Berlin (Ed.), Archäologie in Berlin und Brandenburg 2012. Theiss, Stuttgart, pp. 25 – 29.

Jahns, S., Alsleben, A., Bittmann, F., Brande, A., Christiansen, J., Dannath, Y., Effenberger, H., Giesecke, T., Jäger, K.-D., Kirleis, W., Klooβ, S., Kloss, K., Kroll, H., Lange, E., Medović, A., Neef, R., Stika, H.-P., Sudhaus, D., Wiethold, J., Wolters, S., 2018. Zur Geschichte der nacheiszeitlichen Umwelt und der Kulturpflanzen im Land Brandenburg. Beiträge zur Archäozoologie und Prähistorischen Anthropologie. Langenweisbach, Beier & Beran, pp. 9 – 35.

Kirleis, W., Jahns, S., Dannath, Y., Neef, R., (forthcoming). Früher Ackerbau an der Peripherie - Die Pflanzenfunde linienbandkeramischer und trichterbecherzeitlicher Fundplätze im Havelland, Brandenburg.

Rösch, M., 1999. Botanische Untersuchungen in der bandkeramischen Siedlung, in: Krause, R. (Ed.) Die bandkeramischen Siedlungsgrabungen bei Vaihingen an der Enz, Kreis Ludwigsburg (Baden-Württemberg). Ein Vorbericht zu den Grabungen von 1994–1997. Bericht der Römisch-Germanischen Kommission 79, 1998, 64 – 73.

Uhl, U., 2008. Bandkeramische Siedlungskammer. Neuer Hinweis auf jungsteinzeitliche Bauern bei Lietzow, in: Archäologische Gesellschaft in Berlin und Brandenburg e. V. in Zusammenarbeit mit dem Brandenburgischen Landesmuseum für Ur- und Frühgeschichte und dem Landesdenkmalamt Berlin (Ed.), Archäologie in Berlin und Brandenburg 2007. Theiss, Stuttgart, pp. 61 – 63.

Wolters, S., 2016. Die pflanzlichen Makroreste der Mesolithstation Friesack, in: Benecke, N., Gramsch, B., Jahns, S. (Eds.), Subsistenz und Umwelt der Feuchtbodenstation Friesack 4 im Havelland. Arbeitsberichte zur Bodendenkmalpflege in Brandenburg 29, 189 – 202.

Plate 1. Lietzow 2. 1 and 2 Charred emmer grains, *Triticum dicoccon*, ventral and dorsal views. 3 Charred free-threshing wheat grain, *Triticum aestivum* s.l., ventral, dorsal, and lateral views. 4 Charred pea cotyledon, *Pisum sativum*, outer and inner views. 5 Charred nutshell fragment of hazel, *Corylus avellana*.

Plate 2. Lietzow 2. 1 Charred seeds of lady's thumb, *Persicaria maculosa*. 2 Mass find of charred seeds of black bindweed, *Fallopia convolvulus*. 3 Charred seeds of black bindweed. 4 Charred mouse dropping.

Plate 3. Lietzow 10. 1 Charred emmer grains, *Triticum dicoccon*. 2 and 3 Charred emmer grains, dorsal and lateral views. 4 Charred emmer grain, dorsal, lateral, and ventral views.

Plate 4. Lietzow 10. 1 Charred barren brome grains, *Bromus sterilis*. 2 Single charred barren brome grain, dorsal and ventral views. 3 Clover seeds, cf. *Melilotus*, big-size seeds. 4 Clover seeds, medium-size seeds. 5 Clover seeds, small-size seeds.

7 Southern Denmark: The island of Funen

The island of Funen and the Funen archipelago in their present form are a very young land mass in terms of geological history. It was not until the Quaternary that the island was formed. During the Weichsel glaciation, Funen was transgressed by at least two ice streams. The large glacier tongue of the Baltic Sea was divided into two main branches; one moved through the Little Belt depression west of Funen and the other through the Great Belt depression in the east. In the meantime, Funen was covered by a huge dead ice sheet, the remnant of an earlier north-east ice stream. With the retraction of the ice sheet, a diversified landscape with dynamic geomorphology developed that attracted humans. The ancient glacial landscape of Funen consists of moraine areas; hilly edge moraines with clusters of drumlins, i. e. the elongated, rounded hills that indicate the direction of the former ice movement; and round hat hills. In the south and south-west, Funen is characterized by relatively high, undulating

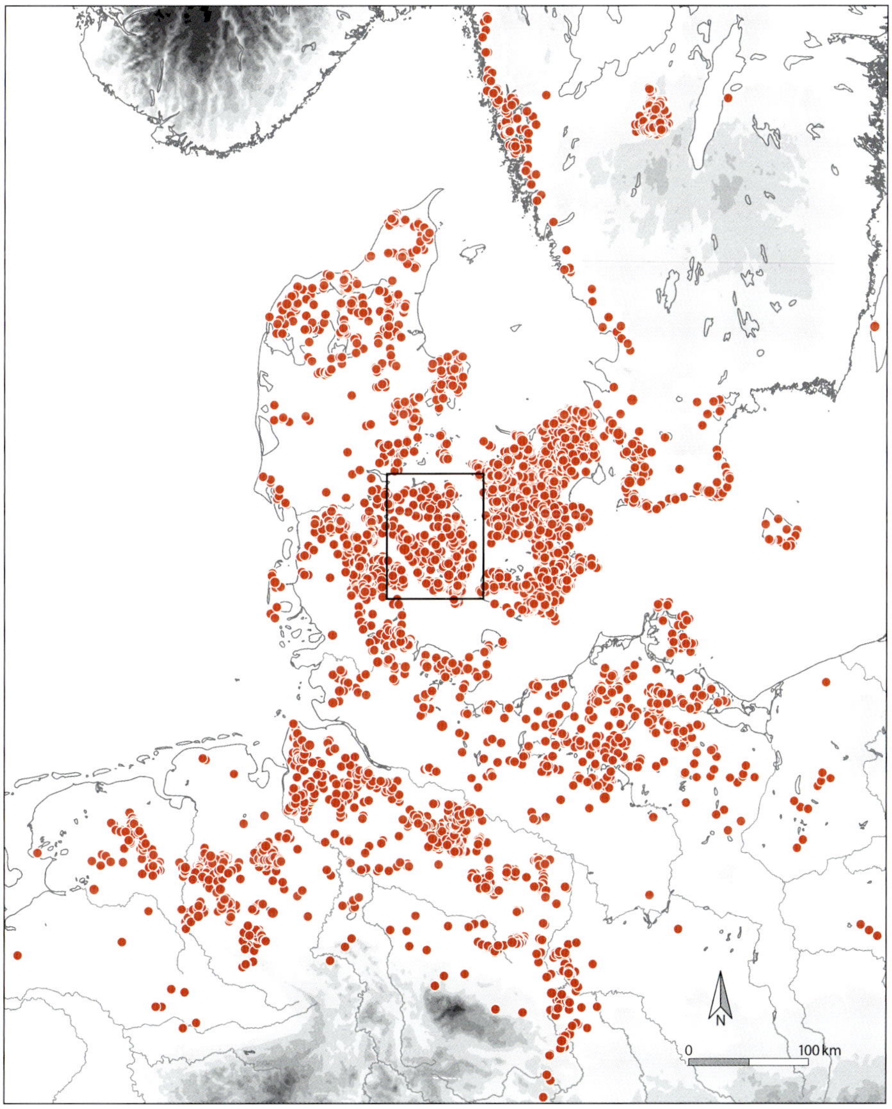

Fig. 7.1. Funen, marked by a frame, located in the core area of the Funnel Beaker Culture based on the distribution of megalithic tombs (after Müller et al., 2019: Fig. 2).

edge moraines and table hill areas, while towards the north rather flat and deeper-lying ground moraines are dominant. Fertile loamy moraine soils are widespread. South-east of the Odense Fjord and in front of the former ice sheet, the soil is somewhat sandier, with sander depositions of the outwash plain (Smed, 1962, 1978). Two palaeo-ecological records are available that allow for insight into the climate history and vegetation development. A sequence from Lake Sarup, in the south-west, highlights a climate-driven change of the water level, coinciding with the 8.2 ka BP cooling event as recorded in the Greenland ice cores that could be deduced from a laminated, high-temporal-resolution part of the core (Bjerring et al., 2013). Pollen and plant macrofossil analysis of an 11 m sediment core from Dallund Sø, a lake situated in the plains in the north of Funen, provide a record of the past 7000 years of vegetation history in response to changing land-use practices. With respect to the Neolithic, around 3900 BCE, a distinct *Ulmus* decline is observed, initiating a shift in composition of the mixed oak woodland. The Neolithic economy during this early stage is though to be primarily based on small-scale arable farming and animal husbandry. A landnam phase—with an increase in woodland opening,

Fig. 7.2. Archaeological sites in the Sarup area, south-western Funen. Red circles = megalithic monuments; brown squares = domestic sites. Contour interval: 2.5 m (after Andersen, 2019: Fig. 2).

the light-demanding hazel, and indicators for human activities—accompanies socio-environmental changes that are expressed through the erection of numerous megalithic tombs (Rasmussen, 2005).

During the Neolithic, Funen island was part of the core area of distribution of the Funnel Beaker groups, which spans from the eastern Jutland peninsula across the south-western Danish archipelago towards Scania, as indicated by a high density of megalithic tombs (Fig. 7.1; Müller, 2019). The identity of the Funnel Beaker groups as farmers can be inferred from features and tools with distinct connections to crop cultivation, such as ard marks. One of the oldest dates for ard marks in northern Europe, dated to c. 3700 cal BCE (4900 ± 40 BP, POZ-28068), comes from the south of Funen, where such structures were found below a long barrow at Højensvej 7, west of Svendborg (Beck, 2013: 70; Sørensen and Karg, 2014).

Intensive archaeological studies were carried out in the south-west of the island of Funen between 1971 and 2014, in the Sarup area, a sandy heathland area covering about 20 km^2 around the village of Sarup (Andersen, 1997, 1999 a, 1999 b, 2018), producing a large body of new data on settlement patterns and ritual landscapes of the Funnel Beaker group. Archaeological survey in the area revealed traces of 88 settlements and more than 120 demolished megalithic structures. The investigations were centred around the discovery and excavation of a two-phased causewayed enclosure dating to the second half of the 4th millennium BCE (Fig. 7.2). The archaeobotanical assemblage of the famous enclosure site of Sarup I (section 7.1) displays high proportions of cereal grains, usually interpreted as being indicative for domestic sites, but thousands of nutshell fragments of hazel show that plant gathering activities were carried out as well. The Frydenlund site (section 7.2) is primarily domestic in character, although the house structures were replaced in a later phase by a type of burial monument of the Barkær type, referring to the long barrows of earth and wood investigated at Barkær, in eastern Jutland. Damsbo FSM 3257 (section 7.3) also has complex stratigraphy, with domestic features overlain by different burial monuments. The domestic site of Sneglebjerg (section 7.4) shows the emmer and barley assemblages that are the typical Funnel Beaker cereals.

Literature:

Andersen, N.H., 1997. Sarup Vol. 1: The Sarup Enclosures: the Funnel Beaker Culture of the Sarup Site, Including Two Causewayed Camps Compared to the Contemporary Settlements in the Area and Other European Enclosures. Aarhus University Press, Aarhus.

Andersen, N.H., 1999a. Sarup Vol. 2: Saruppladsen: Tekst. Aarhus University Press, Aarhus.

Andersen, N.H., 1999b. Sarup Vol. 3: Saruppladsen: Katalog. Aarhus University Press, Aarhus.

Andersen, N.H., 2018. Sarup revisited: archaeological reality and realities of archaeology. Acta Archaeologica 89, 31–62.

Andersen, N.H., 2019. The first farmers as monument builders in the Sarup area, Funen, Denmark, in: Müller, J., Hinz, M., Wunderlich, M. (Eds.), Megaliths, Societies, Landscapes: Early Monumentality and Social Differentiation in Neolithic Europe. Habelt, Bonn, pp. 169–192.

Beck, M., 2013. Højensvej Høj 7—en tidligneolitisk langhøj med flere faser. Aarbøger for nordisk oldkyndighed og Historie 2011–2012, 33–117.

Bjerring, R., Olsen, J., Jeppesen, E., Buchardt, B., Heinemeier, J., McGowan, S., Leavitt, P. R., Enevold, R., Odgaard, B.V., 2013. Climate-driven changes in water level: a decadal scale multiproxy study recording the 8.2-ka event and ecosystem responses in Lake Sarup, Denmark. Journal of Palaeolimnology 49, 267–285.

Müller, J., 2019. Boom and bust, hierarchy and balance: from landscape to social meaning. Megaliths and societies in northern central Europe, in: Müller, J., Hinz, M., Wunderlich, M. (Eds.), Megaliths, Societies, Landscapes: Early Monumentality and Social Differentiation in Neolithic Europe. Habelt, Bonn, pp. 31–77.

Rasmussen, P., 2005. Mid- to late-Holocene land-use change and lake development at Dallund Sø, Denmark: vegetation and land-use history inferred from pollen data. The Holocene 15(8), 1116–1129.

Smed, P., 1962. Studier over den fynske øgruppes glaciale landskabsformer. Meddelelser fra Danske Geologiske Forening 15(1), 1–74.

Smed, P., 1978. Der Fynske Landskab. Copenhagen, Geografforlaget.

Sørensen, L., Karg, S., 2014. The expansion of agrarian societies towards the north—new evidence for agriculture during the Mesolithic/Neolithic transition in southern Scandinavia. Journal of Archaeological Science 51, 98–114.

7.1 Sarup FSM 50

Site name	Sarup FSM 50
Longitude, latitude	55.209032, 10.094958
Federal state, country	Denmark
Excavation campaign	1971–1984
Excavation type	Research excavation
Responsible archaeologist	Niels H. Andersen, Moesgaard Museum, Denmark
Archaeological context	Causewayed enclosure
Analysed structures	Pits, cultural layers, ditches
Archaeological group	Funnel Beaker Culture
Archaeological periodisation	Early to Middle Neolithic (EN II – MN V)
Absolute dating	3400 – 2850 cal BCE
Archaeobotanists	Elske Fischer, Wiebke Kirleis, Kiel (SPP1400); Grethe Jørgensen, Copenhagen
Archaeobotanical sampling strategy	Judgement sampling from archaeological contexts
Analysed sediment volume	923 l
Find concentrations (finds/l)	93
Number of carpological remains	85696

The site of Sarup, in the south-west of the island of Funen (Fig. 7.2), was repeatedly visited in Neolithic times and thus allows the taking of a diachronic perspective (Andersen, 1997, 2004, 2018, 2019). It comprises a temporal chain of two enclosure structures, Sarup I (c. 3400 BCE) and Sarup II (c. 3200 BCE), and of three settlements, Sarup III (c. 3050 BCE), Sarup IV (c. 3000 BCE), and Sarup V (c. 2850 BCE). The sites were erected on a sandy promontory of 9 ha that is located between two watercourses (Fig. 7.1.1). The promontory forms part of a heathland area of 3 × 5 km that is surrounded by more hilly terrain reaching a height of 40 m above sea level. The soils around Sarup are dominated by sand and gravel. The microclimate of the flat heathland area is warmer and dryer than that of the surrounding areas and hence was intensively farmed for several centuries.

Sarup I was constructed during the Early Neolithic II Fuchsberg phase. The enclosure covers an elongated area of about 8 ha. It consists of a palisade fence, fenced enclosures built outside of the palisade, a fenced entrance passage, and two parallel rows of system ditches. Numerous activities took place at the site, including re-cutting and back-filling of the ditches as well as the placement of special assemblages of finds, such as remains of humans and cattle, pottery vessels, axes, grinding stones, querns, and grain. A total of 67 % of the archaeological finds consist of ceramic sherds. The enclosure Sarup II was erected in the Middle Neolithic I b. The structures and finds categories of the enclosure resemble those of Sarup I but cover only half of the area. In the southern part of the promontory, the settlement site of

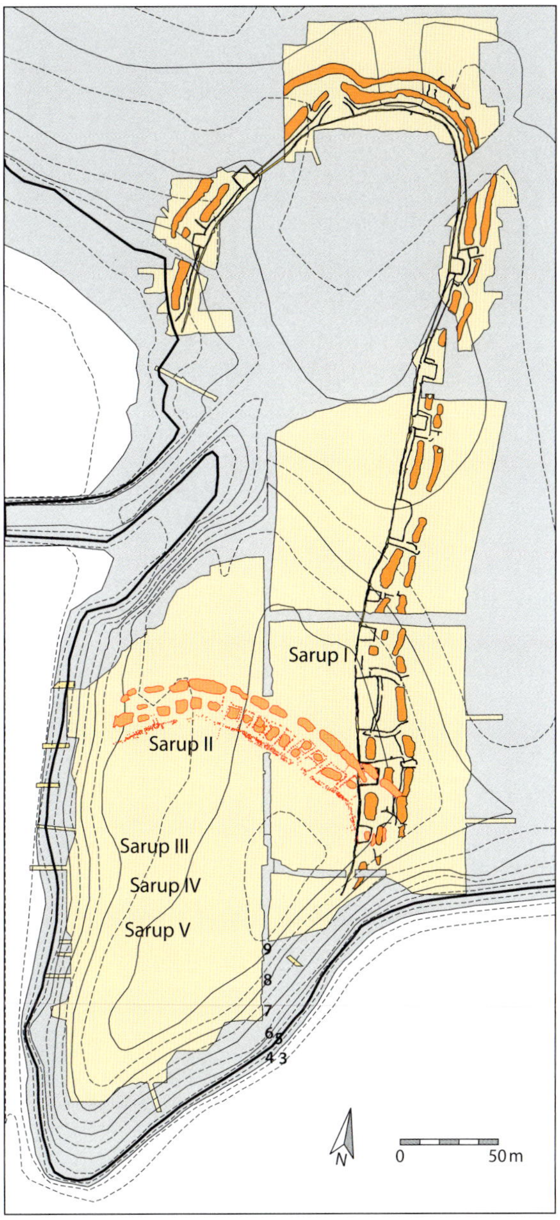

Fig. 7.1.1. The Neolithic sites at Sarup: the enclosures Sarup I and II and the domestic sites Sarup III, IV, and V (after Andersen, 2019: Fig. 9).

Sarup III was discovered, which extended over 4 ha and dates to the Middle Neolithic II. A range of activities was carried out at the site. Postholes and daub indicate possible house structures. Flintknapping, pottery production, and food processing is evidenced by the tens of thousands of ceramic sherds and flint artefacts, numerous stone tools, and animal bones that dominate the archaeological finds spectrum. Another settlement in the centre of the promontory, Sarup IV, covering 4.5 ha, dates to the Middle Neolithic III and IV and is also rich in finds, in particular flint waste and flint with use wear relating to the processing of hides and wood. The structures comprise ritual and storage pits, postholes, and cultural layers. Sarup V dates to the Store Valby phase, or Middle Neolithic V, and, in contrast to Sarup III and Sarup IV, was occupied for only a short time. The main features are pits and cultural layers.

The archaeobotanical material of the Sarup sites is rich, and the overall finds concentration is high (Table 6). The Early Neolithic II enclosure, Sarup I, revealed several mass finds of emmer, two of which also contained numerous caryopses of timothy grass *Phleum pratense* (Jørgensen, 1976, 1982).

Fig. 7.1.2. Collecting windthrow crab apples close to Odderade, Schleswig-Holstein (photo: Wiebke Kirleis, UFG Kiel, October 2018).

Fig. 7.1.3. Windthrow crab apples at the woodland edge close to Odderade in Schleswig-Holstein (photo: Wiebke Kirleis, UFG Kiel, October 2018).

Free-threshing barley and einkorn are present but of minor importance. Of special interest from an archaeobotanical point of view is the deposition of a funnel jar of 42 cm height in a pit in the inner area of the enclosure containing a fragmented beaker and 0.5 l charred emmer grains. Interestingly, emmer grains from the mass finds at Sarup show low cereal $^{\delta15}$N values (< 3‰), which are indicative of long-term, unmanured cultivation (Bogaard et al., 2013). The high number of caryopses of small-seeded Poaceae, in particular timothy grass, indicate extensive cultivation. In the enclosure Sarup II, dating to the Middle Neolithic II phase, emmer remains the dominant cereal, accompanied by single grains of free-threshing barley and free-threshing wheat. Among the weeds, the perennial timothy grass and curlytop knotweed are present. Crab apple remains and nutshells of hazel are recorded in high numbers and show the relevance of plant gathering in the Middle Neolithic II phase. Both of these light-demanding species preferably grow at woodland borders or clearings. They benefitted from the increasing openness of the woodland that was triggered by the Funnel Beaker farmers. Crab apple is an interesting supplement to the regular Neolithic diet that adds a sweet and sour fruit taste. To ensure high yields of crap apple, it is important to prevent domestic and wild animals from browsing on the sweet twigs and the fruits. Crab apple is the only thornless Pomoideae species, and the only one that does not propagate by producing root suckers. Thus it is reasonable to assume that individual trees were carefully managed or, even, that crab apple groves were maintained (section 5.1). The easiest way of harvesting the tiny fruits is to collect the ripe apples from the ground after they have fallen off the tree (Figs. 7.1.2 and 7.1.3). The find of a charred seed of bur-reed *Sparganium* sp. shows that the wetlands in the vicinity of the site were visited and plant material from them was brought into the enclosure. The preference for emmer continues in the settlement site of Sarup III, during the Middle Neolithic III–IV. Barley, here evidenced by grains and single finds of rachis segments, and einkorn were presumably welcome supplements, as was bramble. The weed spectrum comprises black bindweed, tiny Poaceae, and sheep sorrel, the latter possibly originating from plots on poor, sandy soils. The sedge may indicate the proximity of wetlands. The plant assemblage of the Middle Neolithic V settlement of Sarup V is hardly representative. Only very few grains, of free-threshing barley, were identified.

Literature:

Andersen, N.H., 1997. Sarup Vol. 1: The Sarup Enclosures: the Funnel Beaker Culture of the Sarup Site, Including Two Causewayed Camps Compared to the Contemporary Settlements in the Area and Other European Enclosures. Aarhus University Press, Aarhus.

Andersen, N.H., 2004. Sarup: causewayed enclosures placed in a Neolithic ritual landscape on Funen, Denmark. Journal of Nordic Archaeological Science 14, 11–17.

Andersen, N.H., 2018. Sarup revisited: archaeological reality and realities of archaeology. Acta Archaeologica 89, 31–62.

Andersen, N.H., 2019. The first farmers as monument builders in the Sarup area, Funen, Denmark, in: Müller, J., Hinz, M., Wunderlich, M. (Eds.), Megaliths, Societies, Landscapes: Early Monumentality and Social Differentiation in Neolithic Europe. Habelt, Bonn, pp. 169–192.

Bogaard, A., Fraser, R., Heaton, T.H., Wallace, M., Vaiglova, P., Charles, M., Jones, G., Evershed, R.P., Styring, A.K., Andersen, N.H., Arbogast, R.M., Bartosiewicz, L., Gardeisen, A., Kanstrup, M., Maier, U., Marinova, E., Ninov, L., Schafer, M., Stephan, E., 2013. Crop manuring and intensive land management by Europe's first farmers. PNAS 110, 12589–12594.

Jørgensen, G., 1977. Et kornfund fra Sarup: bidrag til tragtbægerkulturens agerbrug. KUML 25, 1976, 47–64.

Jørgensen, G., 1982. Korn fra Sarup: med nogle bemærkninger om agerbruget i yngre stenalder i Danmark. KUML 30, 1981, 221–231.

Plate 1. Sarup FSM 50, Early Neolithic II. 1 Charred emmer grains, *Triticum dicoccon*. 2 Charred grains of free-threshing barley, *Hordeum vulgare nudum*. 3 Charred einkorn grain, *Triticum monococcum*, lateral view. 4 Charred einkorn grain, dorsal and lateral views.

Plate 2. Sarup FSM 50, Middle Neolithic II. 1 Mass find of charred emmer grains, *Triticum dicoccon*. 2 Charred emmer glume bases.

Plate 3. Sarup FSM 50, Middle Neolithic II. 1 Charred barley grains, *Hordeum vulgare*. 2 – 4 Charred grains of free-threshing wheat, *Triticum aestivum/durum*, dorsal and lateral views. 5 Charred grain of free-threshing wheat, dorsal and ventral views.

Plate 4. Sarup FSM 50, Middle Neolithic II. 1 Charred crab apple core, *Malus sylvestris*. 2 – 3 Charred crab apple seeds. 4 Charred flesh of crab apple.

Plate 5. Sarup FSM 50, Middle Neolithic II. 1–2 Charred nutshell fragments of hazel, *Corlyus avellana*.

Plate 6. Sarup FSM 50, Middle Neolithic II. 1 Charred seeds of curlytop knotweed, *Persicaria lapathifolia*. 2 Charred caryopses of timothy grass, *Phleum pratense*. 3 Charred seed of bur-reed, *Sparganium* sp., dorsal and lateral views.

Plate 7. Sarup FSM 50, Middle Neolithic III–IV. 1 Charred emmer grain, *Triticum dicoccon*, dorsal and lateral views. 2 Charred spikelet fork of emmer, front and lateral views. 3 – 6 Charred einkorn grains, *Triticum monococcum*, dorsal and lateral views.

Plate 8. Sarup FSM 50, Middle Neolithic III–IV. 1–3 Charred rachis segments of barley, *Hordeum vulgare*. 4 Charred bramble seed, *Rubus fruticosus*. 5 Charred nutlet of sedge, *Carex* sp. 6 Tiny amber fragment. 7 Charred seeds of black bindweed, *Fallopia convolvulus*.

Plate 9. Sarup FSM 50, Middle Neolithic III–IV. 1 Tiny charred Poaceae, timothy grass and annual meadow grass, *Phleum pratense* and *Poa annua*. 2 Zoomed in: single charred caryopsis of timothy grass. 3 Charred seeds of common sheep sorrel, *Rumex acetosella*.

Plate 10. Sarup FSM 50, Middle Neolithic V. 1 and 2 Charred grains of free-threshing barley, *Hordeum vulgare nudum*, ventral and dorsal views. 3 Charred grain of free-threshing barley, ventral, dorsal and lateral views.

7.2 Frydenlund OBM 6973

Site name	Frydenlund OBM 6973
Longitude, latitude	55.206680, 10.128367
Federal state, country	Denmark
Excavation campaign	2009–2014
Excavation type	Research excavation
Responsible archaeologist	Niels H. Andersen, Moesgaard Museum, Denmark
Archaeological contexts	Settlement site, burials
Analysed structures	Pits, postholes, cultural layers
Archaeological group	Funnel Beaker Culture
Archaeological periodisation	Early Neolithic (EN Ib)
Absolute dating	c. 3650 cal BCE
Archaeobotanists	Elske Fischer, Wiebke Kirleis, Kiel (SPP1400); Marianne Andreasen, Welmoed Out, Moesgaard Museum, Denmark
Archaeobotanical sampling strategy	Systematic sampling of all features and layers
Analysed sediment volume	3167 l
Find concentrations (finds/l)	2
Number of carpological remains	7735

In the south-west of the island of Funen, in Denmark, about 2 km east of the village of Sarup, lies the archaeological site of Frydenlund OBM 6973 (Fig. 7.2). With a radiocarbon age of c. 3650 cal BCE, the Frydenlund site has the earliest traces of Neolithic houses in the Sarup area, and these predate the enclosure site Sarup I. The site was recurrently revisited. The Early Neolithic settlement was followed by non-megalithic and megalithic burial monuments, including a long mound of the Barkær type, which were themselves covered by a Bronze Age barrow (Fig. 7.2.1). The two house structures are marked by postholes. In and around the house sites is a refuse layer containing sherds of numerous pottery vessels and an extensive flint assemblage. The two-aisled house structure (A) in the north has a floor area of 106 m². The structure consists of a row of central postholes and small stake holes around the presumed course of the walls. The latter most probably were wattle-and-daub constructions. Small pits and a thin cultural layer in the west contained numerous charred plant remains, including of durum wheat or rivet wheat, *Triticum durum/turgidum*. In the central part, the deposition of a thin-butted flint axe, ceramic sherds of a Funnel-necked beaker, and amber beads are interpreted as possible grave goods. All the structures were subsequently covered with a carpet of stones. The southern structure (B) also comprises postholes that mark a two-aisled house structure, in this case covering an area of 64 m² and with only few finds and features. After removal of two roof-bearing posts, a plank-coffin burial and a small stone covering were placed in its centre. All the southern features (= structure B) were surrounded by an even later fence structure. Two façade ditches found in the west of each structure are part of the Barkær structures. Both function as foundations for the erection of three massive wooden planks that were split off a trunk of about 1.7 m diameter. The planks were supported by numerous stones and by several boulders weighing about 50 kg each. Given the deep and stable foundation, the planks are thought to have reached up to 7.5 m into the sky and were an impressive construction (Andersen, 2015, 2018, 2019).

Archaeobotanical investigations on the material from the 2009–2012 excavations were carried out at Kiel in the frame of the SPP1400. The later excavations used archaeobotanists from the Moesgaard Museum. The Early Neolithic settlement traces were dated to EN I b and EN II and up to MN I. The site also yielded individual archaeobotanical finds from the late Funnel Beaker Culture (MN V) and the Single Grave Culture (YN). The Early Neolithic cultural layers were sampled the most extensively. Accordingly, the samples from house structures A and B were particularly rich in finds, the most common being cereal caryopses. Free-threshing wheat was identified mainly in features belonging to structure A, while barley dominates in the other parts of the settlement. Threshing remains are largely limited to the cultural layers of house structure A, in particular to Feature A52. Here, hundreds of rachis segments of free-threshing wheat with tetraploid genomic constitution, that is, durum or rivet wheat, were encountered (Plate 3 and 4). The sheer number of finds suggests that this wheat most probably was grown locally. Finds of rachis segments of free-threshing wheat are a rare but important occurrence. These findings allow for an in-depth archaeobotanical distinction between the free-threshing wheat species with different genomic constitution, that is, the

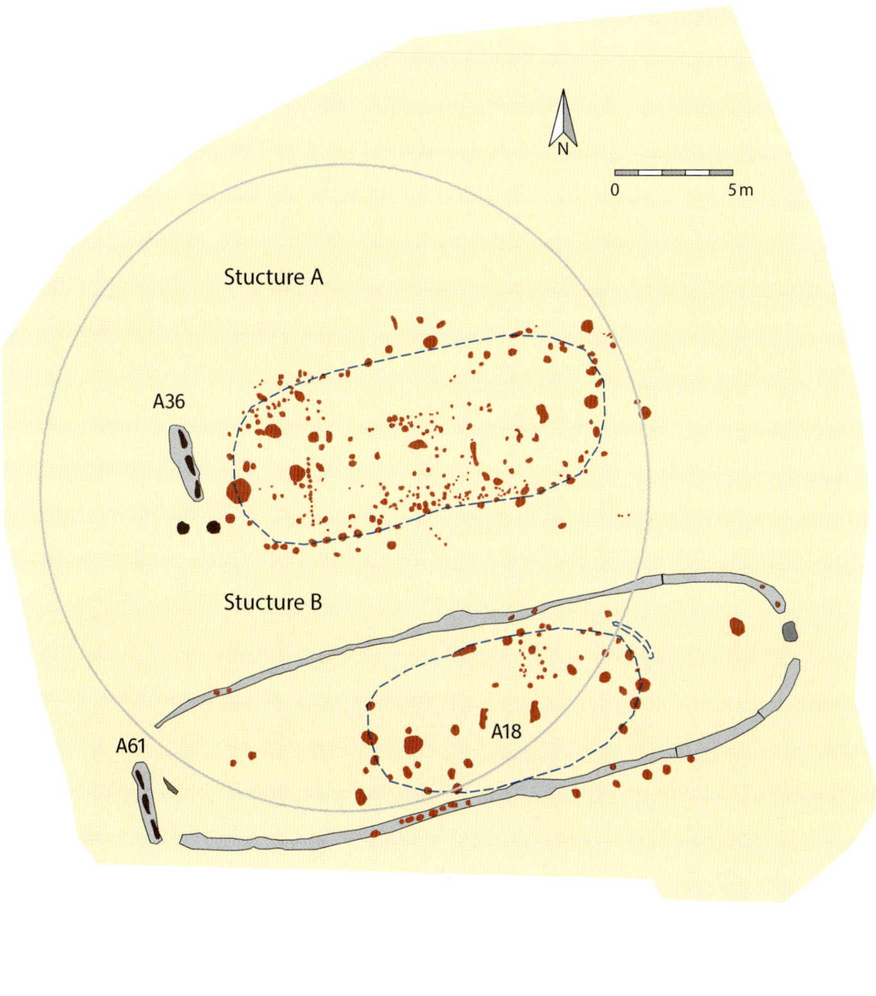

Fig. 7.2.1. Plan of the settlement, Frydenlund. The contours of the EN house structures are shown as dashed blue lines; the postholes are marked in red; the southern structure was framed by a post-built fence, shown in light grey; structures A36 and A61 are the façade ditches with the massive planks, shown in brown; the Early Neolithic structures were subsequently covered by a Bronze Age barrow, shown as a grey circle (after Andersen, 2019: Fig. 3).

Fig. 7.2.2. Sites with archaeobotanical evidence for tetraploid free-threshing wheat in central and northern Europe (after Kirleis and Fischer, 2014: Fig. 4).

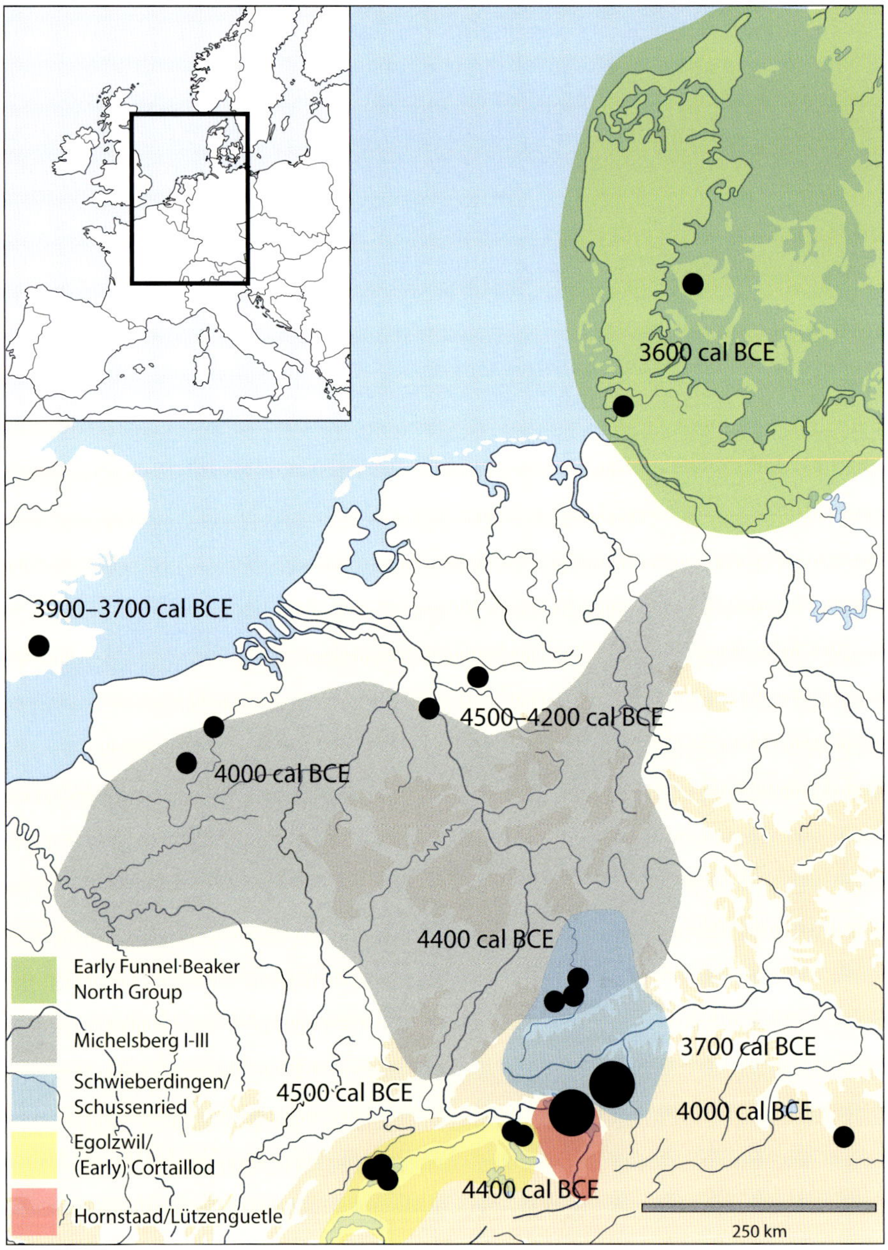

hexaploid bread wheat *Triticum aestivum* and the tetraploid *Triticum durum/ turgidum*, since the rachis segments owe specific morphological traits. This is in contrast to the caryopses, which cannot be differentiated based on their morphology. The traits of the rachis segments that are characteristic of the tetraploid free-threshing wheat species are the nodes that show a rounded lump beneath each glume insertion; the more or less straight-sided trapezoid shape of the rachis segment, with only a slight incurved narrowing immediately below the node; the lack of any longitudinal incision or line on the rachis segments; and the rounded edge of each rachis segment (Jacomet, 2006).

The Frydenlund mass finds constitute the first proof that farmers of the Neolithic Funnel Beaker Culture cultivated free-threshing wheat with tetraploid genomic constitution and thereby indicates their supra-regional communication networks. The first evidence for this species in a Funnel Beaker context comes from the site Albersdorf LA68, Dieksknöll (section 3.4). There, an imprint of rachis segments in a ceramic sherd was found indicating that free-threshing wheat with tetraploid genomic constitution was present and a known species. However, this was not enough evidence to argue for local cultivation. A compilation of the distribution of Neolithic finds of free-threshing wheat with tetraploid genomic constitution clearly shows the south-western distribution of this species (Fig. 7.2.2). The Frydenlund finds of tetraploid free-threshing wheat are thus the missing link that proves the existence of interaction between the Neolithic Michelsberg and the Neolithic Funnel Beaker groups. They add an archaeobotanical perspective to archaeological evidence indicating strong links between these groups, namely, the distribution of specific artefacts and a crossover in lifestyle that is expressed through the construction of causewayed enclosures (Kirleis and Fischer, 2014; Kreuz et al., 2014).

Gathered plants are a regular occurrence in the Frydenlund samples (Table 6). They comprise hazel, bramble, and raspberry. Crab apple, however, is missing. More than 10 different weed species constitute a high diversity, but in terms of numbers they are rare. The most common weed species are small-size clovers, cleavers, vetches, and black bindweed. Vegetative plant remains of lesser celandine and tuber oat grass supplement the assemblages. In particular the cultural layers in structure A contained numerous tubers of lesser celandine (Klooß et al., 2016). These starch-rich tubers may have contributed to the daily nutrition (see also section 7.3). Charred fruits of lime indicate either the use of this tree species or the presence of a lime tree within the settlement area that spread its numerous fruits in late summer/early autumn. Additional material from the Frydenlund excavations in 2013 and 2014 was studied by Marianne Høyem Andreasen (2017). The wood charcoal, analysed by Welmoed Out from the Moesgaard Museum, shows a regular presence of Pomoideae (personal communication, Welmoed Out, 09/2018), that may be interpreted as potentially adding evidence at least for the use of crab apple wood.

Literature:

Andersen, N.H., 2015. Frydenlund, Early Neolithic settlement and "barkær" structures in the Sarup area. Acta Archaeologica Lundensia 8(65), 117–127.

Andersen, N.H., 2018. Sarup revisited: archaeological reality and realities of archaeology. Acta Archaeologica 89, 31–62.

Andersen, N.H., 2019. The first farmers as monument builders in the Sarup area, Funen, Denmark, in: Müller, J., Hinz, M., Wunderlich, M. (Eds.), Megaliths, Societies, Landscapes: Early Monumentality and Social Differentiation in Neolithic Europe. Habelt, Bonn, pp. 169–192.

Høyem Andreasen, M., 2017. FSM 6973, Frydenlund (FHM 4296/1119), Makrofossilanalyse af anlæg fra tidlig-neolitikum. Afdeling for konservering og naturvidenskab, Moesgaard Museum, Højbjerg.

Jacomet, S., 2006. Identification of cereal remains from archaeological sites, fourth ed. Archaeobotany Lab IPAS, Basel University, Basel. https://duw.unibas.ch/fileadmin/user_upload/duw/IPNA/PDF_s/PDF_s_in_use/Cereal_Id_Manual_engl.pdf

Kirleis, W., Fischer, E., 2014. Neolithic cultivation of tetraploid free threshing wheat in Denmark and northern Germany: implications for crop diversity and societal dynamics of the Funnel Beaker Culture. Vegetation History and Archaeobotany 23, Supplement 1, 81–96.

Klooß, S., Fischer, E., Out, W., Kirleis, W., 2016. Charred root tubers of lesser celandine (*Ficaria verna* HUDS.) in plant macro remain assemblages from northern, central and western Europe. Quaternary International 404, 25–42.

Kreuz, A., Märkle, T., Marinova, E., Rösch, M., Schäfer, E., Schamuhn, S., Zerl, T., 2014. The Late Neolithic Michelsberg culture—just ramparts and ditches? A supraregional comparison of agricultural and environmental data. Praehistorische Zeitschrift 89, 72–115.

Plate 1. Frydenlund OBM 6973. 1 Mass find of charred cereal grains. 2 and 3 Charred grains of free-threshing wheat, *Triticum aestivum/durum*.

Plate 2. Frydenlund OBM 6973. 1–4 Charred emmer grains, *Triticum dicoccon*, dorsal, and lateral views. 5 Charred einkorn grain, *Triticum monococcum*, dorsal, and lateral views. 6 Charred grain of free-threshing barley, *Hordeum vulgare nudum*, dorsal, and ventral views.

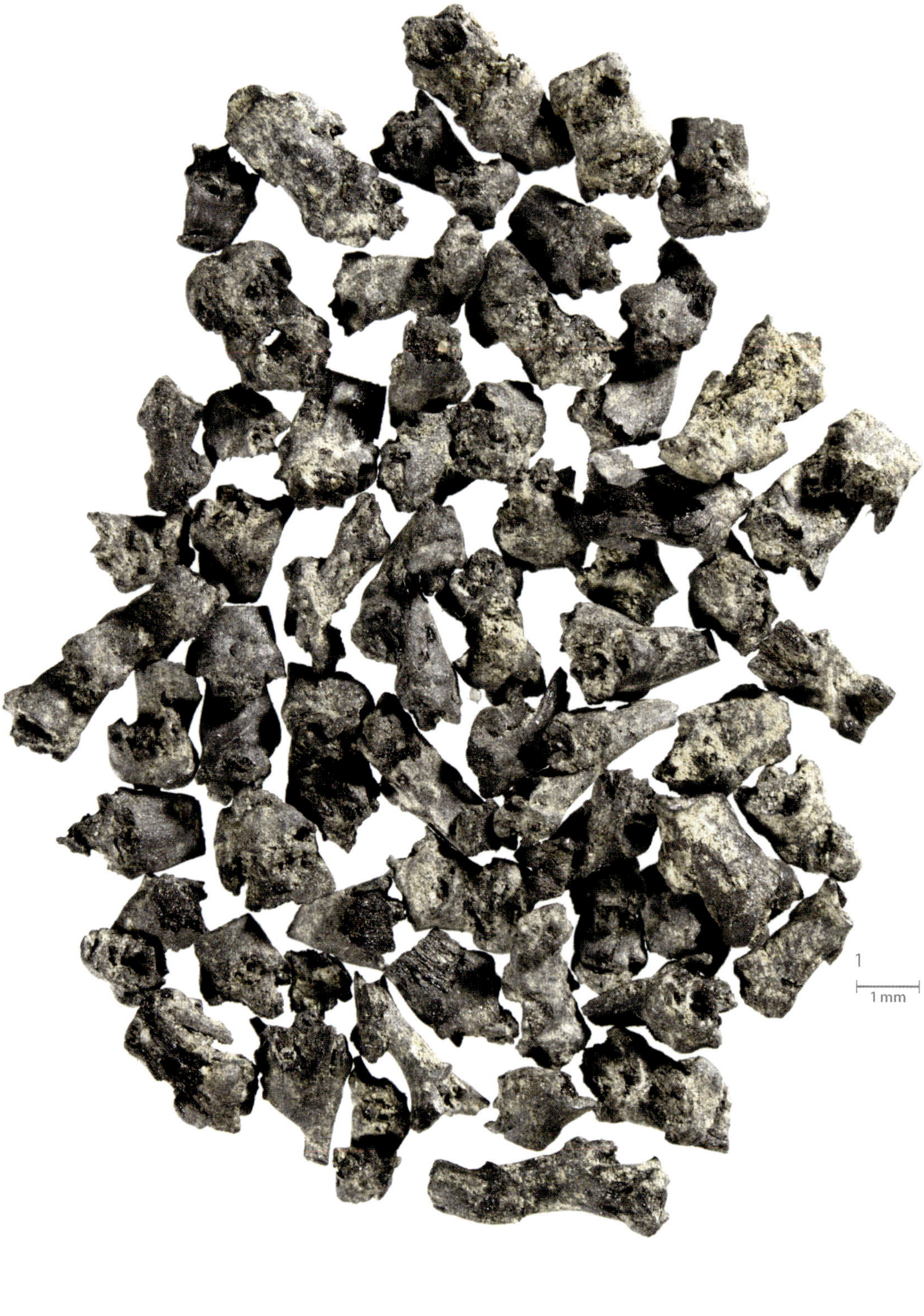

Plate 3. Frydenlund OBM 6973. 1 Charred rachis segments of free-threshing wheat with tetraploid genomic constitution, *Triticum durum/turgidum*.

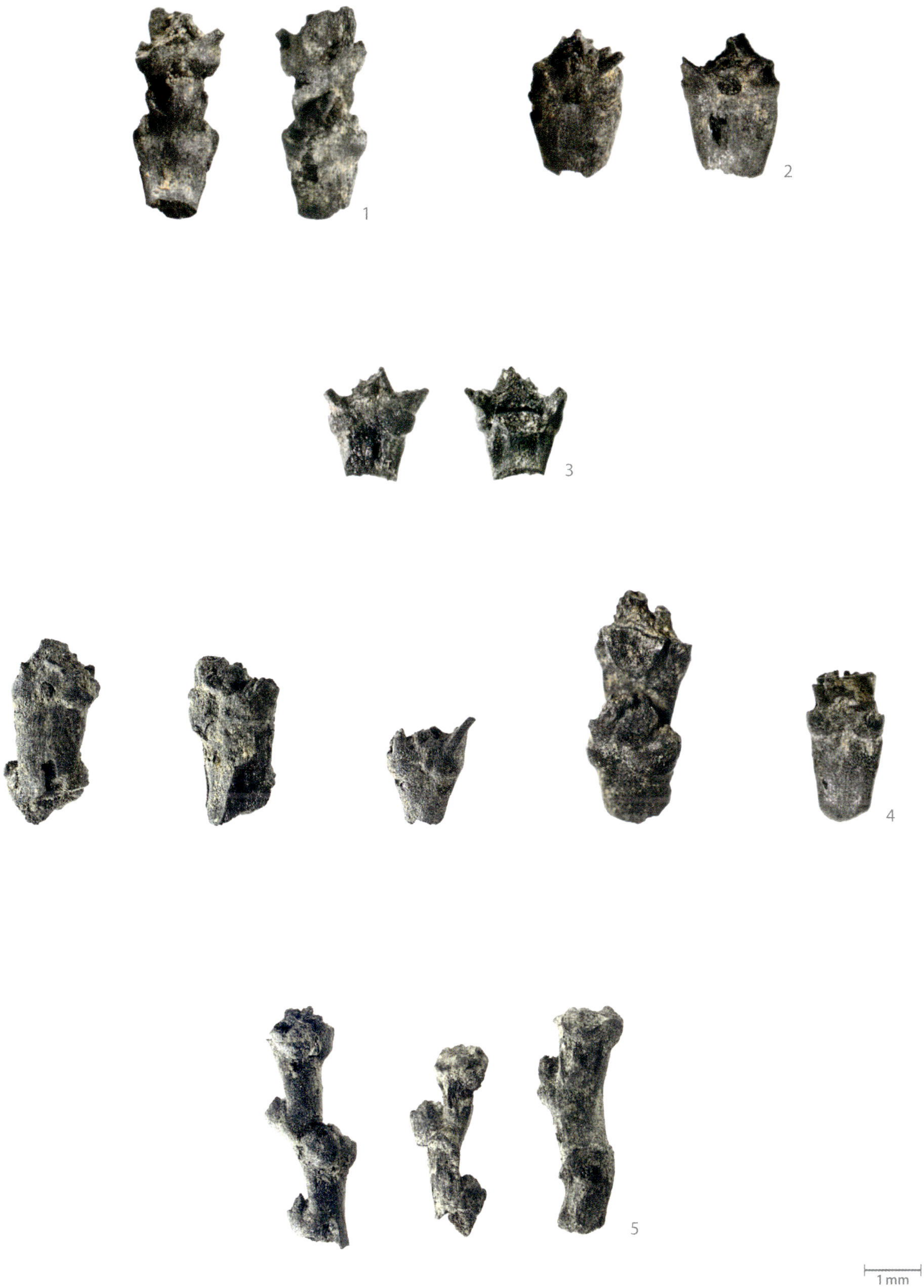

Plate 4. Frydenlund OBM 6973. 1–5 Charred rachis segments of free-threshing wheat with tetraploid genomic constitution, *Triticum durum/turgidum* (photos: Anna E. Reuter, UFG Kiel).

Plate 5. Frydenlund OBM 6973. 1 Charred glume bases of emmer and einkorn, *Triticum monococcum/dicoccon*. 2 Charred fragment of hazel nutshell, *Corylus avellana,* outer and inner views. 3 Charred nutshell fragment of hazel. 4 Charred bramble seed, *Rubus fruticosus*.

Plate 6. Frydenlund OBM 6973. 1 and 2 Charred tubers of lesser celandine, *Ficaria verna*, variable sizes and shapes (photo: Stefanie Klooß, UFG Kiel (2)).

Plate 7. Frydenlund OBM 6973. 1 Charred field brome grain, cf. *Bromus arvensis*, dorsal and ventral views. 2 Seed of yellow clover, cf. *Medicago lupulina*. 3 Charred fruits of lime tree, *Tilia* sp. 4 Charred indeterminable seed. 5 and 6 Charred indeterminable remains.

7.3 Damsbo FSM 3527

Site name	Damsbo FSM 3527
Longitude, latitude	55.166113, 10.133681
Federal state, country	Denmark
Excavation campaign	2002–2008
Excavation type	Research excavation
Responsible archaeologist	Niels H. Andersen, Moesgaard Museum, Denmark
Archaeological contexts	Settlement site, burials
Analysed structures	Cultural layers below megalithic tomb, pits, postholes
Archaeological group	Funnel Beaker Culture; Single Grave Culture
Archaeological periodisation	Early to Younger Neolithic (EN II–YN 1/2)
Absolute dating	3400–2600 cal BCE
Archaeobotanists	Marcin Przybyla, Stefanie Klooß, Wiebke Kirleis, Kiel
Archaeobotanical sampling strategy	Judgement sampling from archaeological contexts
Analysed sediment volume	897 l
Find concentrations (finds/l)	20
Number of carpological remains	17452

The locality of Damsbo is situated on Helnaes Bay, c. 2.5 km south-east of Sarup, in the south-west of the island of Funen (Fig. 7.2). Several sites were excavated here within an area of 200 × 80 m (Fig. 7.3.1; Andersen, 2004, 2008, 2009). In the south, a cluster consisting of a long barrow (A6), two detached dolmens (A5 and A38), and a small passage tomb (A4) was excavated, and in the north, a long barrow covering an Early Neolithic house structure (A2), two detached dolmens (A3 and A32), and a passage grave, again on top of an Early Neolithic house structure (A1). The house structure of A2, with a floor area of 70 m^2, was two-aisled, with five postholes for the large, roof-bearing posts in a row down the middle and smaller ones marking the walls. After the house was burned down, the ground was ploughed over with an ard before the long barrow was erected. The sequence of house abandonment, ploughing, and erection of a long barrow in the same spot is typical for the Funnel Beaker Culture. A number of finds emerged from the area around the house structure which date to the Early Neolithic II, the so-called Fuchsberg phase. The northern long barrow (A121) at the Damsbo site also was erected on top of a two-aisled longhouse, in this case covering an area of approximately 90 m^2 and with six postholes of roof-bearing posts in a row. Both house areas yielded only a limited number of finds. In addition, the features of A121 have been heavily affected by soil erosion (Andersen, 2009: 41, 2011).

The Early Neolithic cultural layers below the megalithic tombs at Damsbo FSM 3527 contained relatively few and poorly preserved botanical macro-remains (Table 6). In particular cereal grains are extremely rare. The plant assemblage is dominated by remains of wild species, such as fragments of nutshells of hazel, seeds of bramble and hawthorn, fruits of lime (linden), and root tubers of lesser celandine. All these wild plants grow in woodland habitats. Their presence thus shows that the settlement most probably was situated in a small clearing in an otherwise still densely forested environment. The inhabitants must have had an intimate relationship with the surrounding woodlands. A sample from posthole A158 comprises a special set of botanical remains, containing 14 bulbs of tuber oat grass, numerous seeds of white goosefoot and sheep sorrel, ribwort plantain, brome, and a cupula fragment of an unripe acorn. This secondary infill containing mainly species indicative for the opening of the landscape may hint to waste disposal following a specific activity. The Early Neolithic assemblages from the domestic contexts show, in general, an astonishingly low number of cereal remains. Different scenarios can explain this fact. The houses may have had a special function that excluded

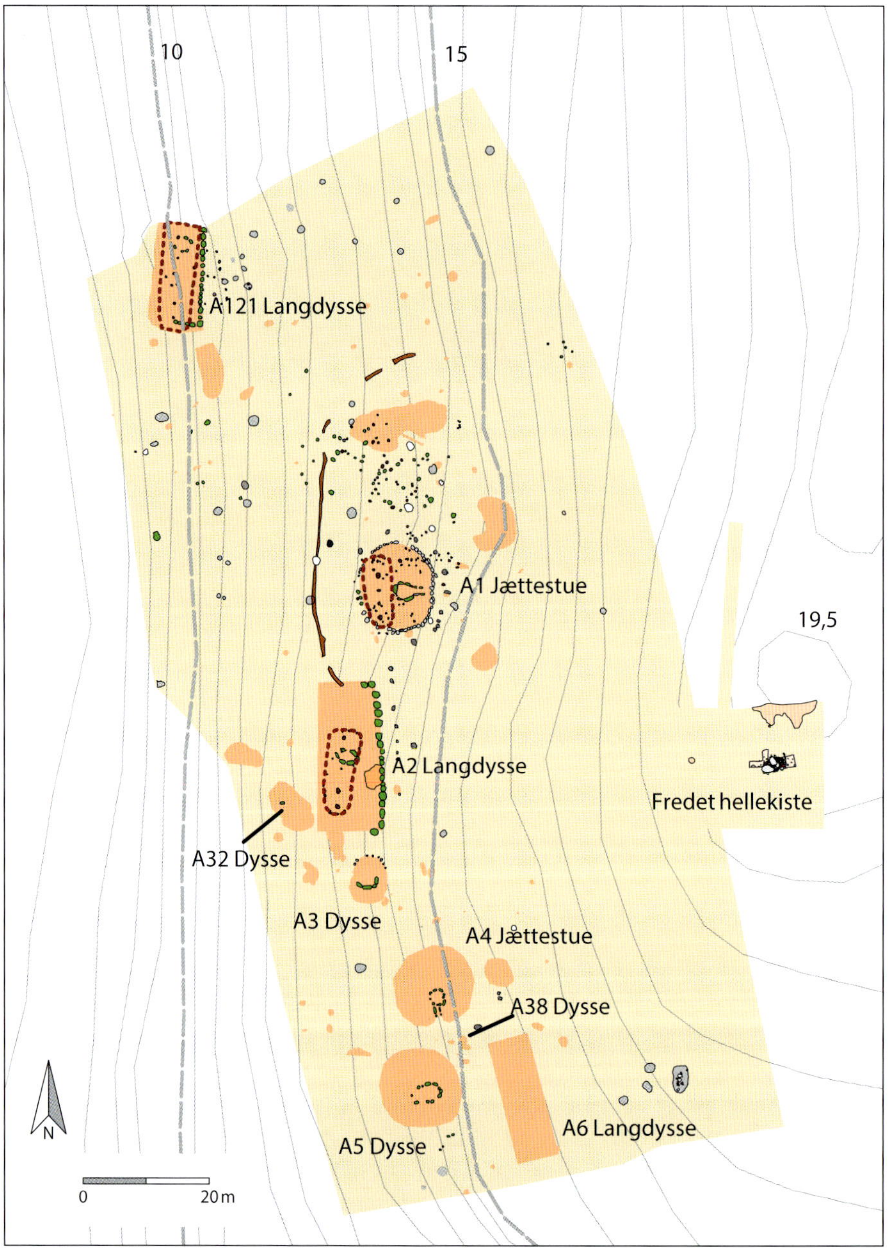

Fig. 7.3.1. Neolithic structures, Damsbo Mark. The postholes marked in red indicate three Early Neolithic two-aisled houses. After abandonment, the areas were ploughed and re-used to erect megalithic monuments (after Andersen, 2019, Fig. 8).

food processing activities and cereal storage, or they may have had a short lifetime, or they may have been intensively cleaned before abandonment and intentional burning (Andersen, 2011: 150).

In contrast to the Early Neolithic two very different mass finds were detected, dating into the first centuries of the 3rd millennium BCE. Pit A98, of Middle Neolithic V Store Valby origin, contained a mass find of emmer glume bases with a minor contribution of emmer and barley grains, combined with masses of (rye) brome caryopses, and further minor admixture of weeds, such as timothy grass and white goosefoot. This combination of threshing remains with weed remains accounts for an interpretation of this assemblage as waste disposal. The Younger Neolithic mass find from pit A30, attributed to the Single Grave Culture, consists of thousands of emmer grains, outnumbered by several thousand emmer glume bases. The emmer again is accompanied by numerous caryopses of rye brome and timothy grass. Nutshell fragments of hazel and fragments of crab apples are also present. It is striking that this emmer mass find has a prominent admixture of tiny and large-size caryopses of wild grasses, namely, (rye) brome and timothy grass. This mass find is also most probably the result of waste disposal, although the tiny grass caryopses and emmer grains constitute a starch-rich edible plant food component.

Literature:

Andersen, N.H., 2004. Sarup: causewayed enclosures placed in a Neolithic ritual landscape on Funen, Denmark. Journal of Nordic Archaeological Science 14, 11–17.

Andersen, N.H., 2008. Rapport over udgravninger på Damsbo´s jorder på Sydvestfyn i årene 2002 til 2008. Moesgaard Museum, Højbjerg.

Andersen, N.H., 2009. The Sarup area in southwestern Funen at the end of the fourth millennium BC, in: Schülke, A. (Ed.), Plads og rum i traegtbaegerkulturen. Kongelige Nordiske Oldskriftselskab, Copenhagen, pp. 38 – 44.

Andersen, N.H., 2011. Causewayed enclosures and megalithic monuments as media for shaping Neolithic identities, in: M. Furholt, Lüth, F., Müller, J. (Eds.), Megaliths and Identities: Early Monuments and Neolithic Societies from the Atlantic to the Baltic. Habelt, Bonn, pp. 143 – 154.

Andersen, N.H., 2019. The first farmers as monument builders in the Sarup area, Funen, Denmark, in: Müller, J., Hinz, M., Wunderlich, M. (Eds.), Megaliths, Societies, Landscapes: Early Monumentality and Social Differentiation in Neolithic Europe. Habelt, Bonn, pp. 169–192.

Plate 1. Damsbo FSM 3527, Early Neolithic II. 1–2 Charred grains of barley, *Hordeum vulgare*, ventral views. 3 Charred grain of Cerealia indet. 4 Charred seeds of hawthorn, *Crataegus* sp. 5 Charred bramble seed, *Rubus fruticosus*. 6 Charred root tubers of lesser celandine, *Ficaria verna*.

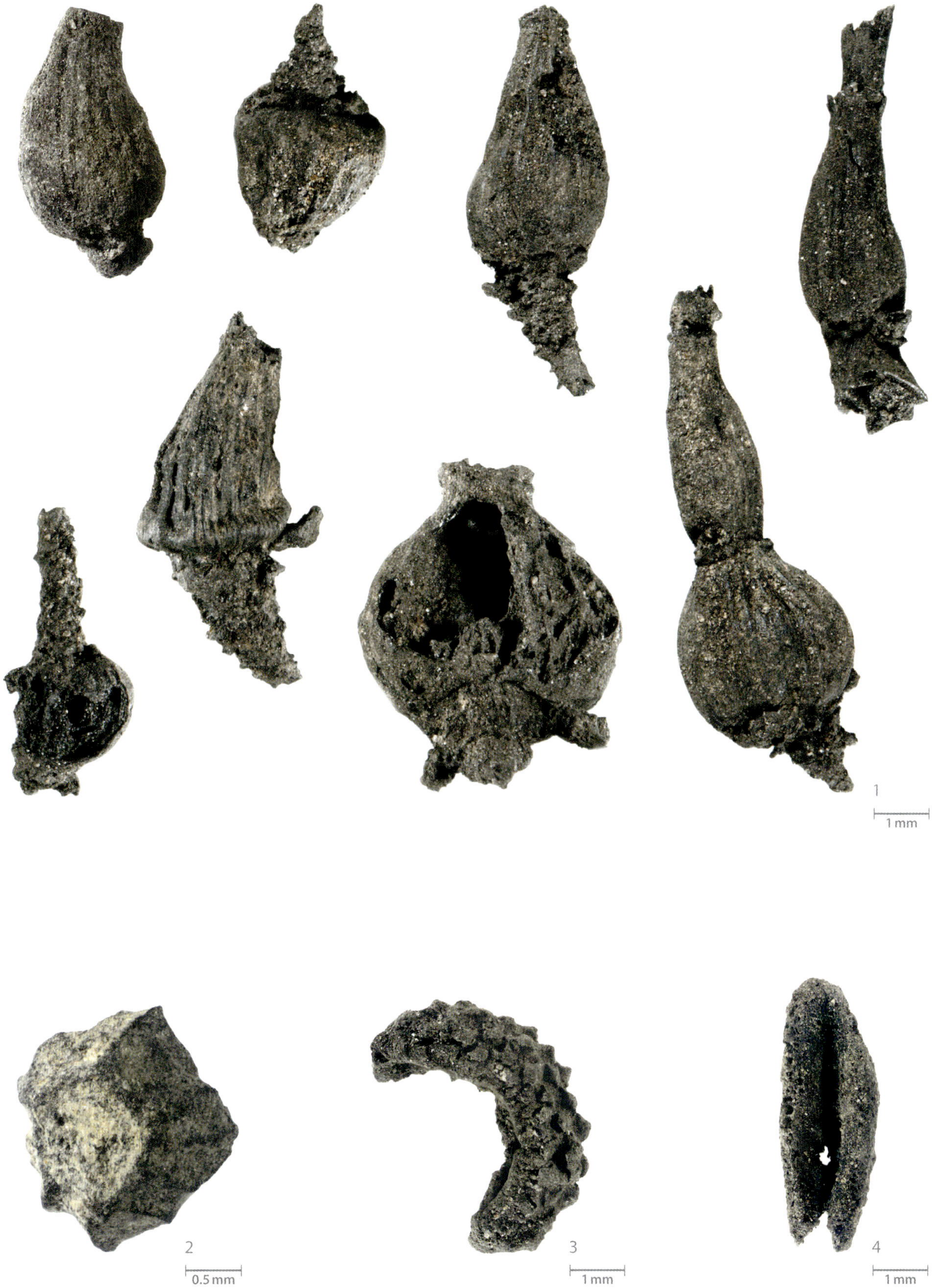

Plate 2. Damsbo FSM 3527, Early Neolithic II. 1 Charred bulbs of tuber oat grass, *Arrhenatherum elatius* var. *bulbosum*. 2 Charred fruit of lime tree, *Tilia* sp. 3 Charred cupula of unripe acorn, *Quercus* sp. 4 Charred brome caryopsis, *Bromus* sp., ventral view.

Plate 3. Damsbo FSM 3527, Early Neolithic II. 1 Charred fruits of black bindweed, *Fallopia convolvulus*. 2 Charred seeds of ribwort plantain, *Plantago lanceolata*. 3 Charred seeds of white goosefoot, *Chenopodium album*. 4 Charred seeds of common sheep sorrel, *Rumex acetosella*.

Plate 4. Damsbo FSM 3527, Middle Neolithic V. 1 Charred barley grains, *Hordeum vulgare*. 2 Charred emmer grains, *Triticum dicoccon*. 3 Charred glume bases of emmer.

Plate 5. Damsbo FSM 3527, Middle Neolithic V. 1 Charred grains of rye brome, *Bromus secalinus*. 2 Charred brome grains, *Bromus* sp.

Plate 6. Damsbo FSM 3527, Middle Neolithic V. 1 Charred caryopses of timothy grass, *Phleum pratense*.
Damsbo FSM 3527, Funnel Beaker period (general). 2 Charred nutshell fragments of hazel, *Corylus avellana*. 3 Charred bramble seeds, *Rubus fruticosus*. 4 Charred fruit of lime tree, *Tilia* sp.

Plate 7. Damsbo FSM 3527, Younger Neolithic. 1 Charred emmer grains, *Triticum dicoccon*. 2 Charred glume bases of emmer.

Plate 8. Damsbo FSM 3527, Younger Neolithic. 1 Charred caryopses of timothy grass, *Phleum pratense*. 2 Charred crab apple core, *Malus sylvestris*. 3 Charred crab apple flesh. 4 Charred crab apple epidermis.

7.4 Sneglebjerg FHM 9137

Site name	Sneglebjerg FHM 9137
Longitude, latitude	55.198921, 10.121722
Federal state, country	Denmark
Excavation campaign	1998–1999
Excavation type	Research excavation
Responsible archaeologist	Niels H. Andersen, Moesgaard Museum, Denmark
Archaeological context	Settlement site
Analysed structures	Cultural layers
Archaeological group	Funnel Beaker Culture
Archaeological periodisation	Early Neolithic-Middle Neolithic (EN II–MN Ib)
Absolute dating	c. 3300 cal BCE
Archaeobotanists	Elske Fischer, Wiebke Kirleis, Kiel (SPP1400)
Archaeobotanical sampling strategy	Judgement sampling from archaeological contexts
Analysed sediment volume	1150 l
Find concentrations (finds/l)	1 (0.2)
Number of carpological remains	220

The site of Sneglebjerg is situated in the south-east of the village of Strandby, Funen (Fig. 7.2). The excavation of a cultural layer of 44 m^2 resulted in the massive amount of c. 3700 ceramic and flint artefacts and more than 800 bones being collected. The site is dated to the transition from Early Neolithic II to Middle Neolithic Ib, with pottery in the Fuchsberg style. The high find concentrations suggest either local production of pottery and flint tools or intentional deposition of refuse. Possibly, waste was collected at other places and then cumulatively disposed of in a special area (Andersen, 2001, 2004, 2009: 42–43).

In contrast to the archaeological artefacts, the archaeobotanical assemblage is rather small, with 218 finds altogether (Table 6). Charred cereal grains dominate, but these are poorly preserved. Therefore just one fifth of the cereal grains could be identified to the genus or species level. The selection of identifiable grains shows a dominance of emmer over barley, the latter most probably being of the free-threshing variety. Only a single emmer glume base was found. The few occurrences of potential weed remains comprise goosefoot and lady's thumb.

Literature:

Andersen, N.H., 2001. Rapport over udgravninger af lokaliteten Sneglebjerg, Udgravning af kulturlag fra mellemste del af tragtbægerkulturens tid ca. 3.300 f. Kr. (FHM 4241, OBM 9137, RAS P.3668/00, Sb. xx af Hårby sogn, Baag herred). Moesgaard Museum, Højbjerg.

Andersen, N.H., 2004. Sarup. Causewayed enclosures placed in a Neolithic ritual landscape on Funen, Denmark. Journal of Nordic Archaeological Science 14, 11–17.

Andersen, N.H., 2009. The Sarup area in southwestern Funen at the end of the fourth millennium BC, in: Schülke, A. (Ed.), Plads og rum I traegtbaegerkulturen. Kongelige Nordiske Oldskriftselskab, Copenhagen, 38–44.

Plate 1. Sneglebjerg FHM 9137. 1 Charred emmer grains, *Triticum dicoccon*, dorsal and lateral views. 2 Charred emmer grains, dorsal, ventral and lateral views. 3 Charred grains of free-threshing barley, *Hordeum vulgare nudum*, dorsal and ventral views.

Plate 2. Sneglebjerg FHM 9137. 1– 3. Charred barley grains, *Hordeum vulgare*, dorsal, ventral and lateral views.

Plate 3. Sneglebjerg FHM 9137. 1 Charred nutshell fragments of hazel, *Corylus avellana*. 2 Charred seeds of goosefoot, *Chenopodium* sp. 3 Charred seeds of lady's thumb, *Persicaria maculosa*.

Appendices

Table 1. Absolute find numbers of charred archaeobotanical remains per site, section 2.

Section no./site name	2.1 Rosdorf-Mühlengrund	2.2 Rullstorf 8	2.3 Hittfeld 61	2.4 Lavenstedt 178	2.5 Sievern 114	2.6 Vechta 10	
Archaeological periodisation	LBK	EN II–MN II	FBC	MN I–II	Neol.	YN–LN	
Crops							
Cereals, caryopses							
Triticum monococcum	421	.	.	.	.	.	Einkorn
Triticum dicoccon	1278	23	50	39	.	33	Emmer
Triticum monococcum/dicoccon	567	.	.	.	.	.	Einkorn/emmer
Triticum aestivum s. l./*durum/turgidum*	2	.	.	4	.	.	Free-threshing wheat
Triticum sp.	4855	14	4	3	.	.	Wheat
Hordeum vulgare nudum	.	.	1	47	.	48	Free-threshing barley
Hordeum vulgare	137	.	13	187	.	24	Barley
Cerealia indet.	20464	28	137	592	.	103	Cereals
Sum cereals, caryopses	**27724**	**65**	**205**	**872**	**0**	**208**	
Cereals, threshing remains							
Triticum monococcum, glume bases	3993	2	.	1	.	.	Einkorn
Triticum dicoccon, threshing remains	3076	250	29	59	.	.	Emmer
Triticum monococcum/dicoccon, glume bases	346	.	.	7	.	.	Einkorn/emmer
Sum cereals, threshing remains	**7415**	**252**	**30**	**67**	**0**	**0**	
Oil plants							
Linum usitatissimum	31	.	.	.	.	.	Linseed
Papaver somniferum	1	.	.	.	.	.	Opium poppy
Sum oil plants	**32**	**0**	**0**	**0**	**0**	**0**	
Legumes							
Lens culinaris	6	1	.	.	.	.	Lentil
Pisum sativum	243	6	.	.	.	.	Garden pea
Fabaceae p. p. (cultivated)	6	7	.	.	.	.	Pulses (cultivated)
Sum legumes	**255**	**14**	**0**	**0**	**0**	**0**	

Table 1 (continued). Absolute find numbers of charred archaeobotanical remains per site, section 2.

Section no./site name	2.1 Rosdorf-Mühlengrund	2.2 Rullstorf 8	2.3 Hittfeld 61	2.4 Lavenstedt 178	2.5 Sievern 114	2.6 Vechta 10	
Archaeological periodisation	LBK	EN II–MN II	FBC	MN I–II	Neol.	YN–LN	
Fruits/nuts/tubers							
Corylus avellana, nutshell fragments	39	39	2	52	5	5	Hazel
Crataegus sp.	.	1	.	.	.	.	Hawthorn
Ficaria verna, root tubers	.	2	.	.	.	.	Lesser celandine
Fragaria vesca	.	1	.	.	.	.	Wild strawberry
Malus sylvestris, seeds	.	.	4	.	.	.	Crab apple
Malus sylvestris, flesh	.	.	4	1	.	.	Crab apple
Quercus sp.	.	.	.	.	.	5	Oak
Rosa sp.	.	.	.	1	.	.	Rose
Rubus fruticosus	.	12	3	1	.	.	Bramble
Rubus idaeus	.	6	1	.	.	.	Raspberry
Rubus sp.	.	1	.	.	.	.	Bramble/Raspberry/Dewberry
Sum fruits/nuts/tubers	**39**	**62**	**14**	**55**	**5**	**10**	
Wild taxa (dryland)							
Ajuga reptans	.	3	.	.	.	.	Carpet bugleweed
Bromus secalinus	.	.	.	1	.	.	Rye brome
Bromus secalinus-type	9	.	.	.	.	.	Rye brome-type
Chenopodium album	140	1	.	8	.	.	White goosefoot
Chenopodium ficifolium	18	.	.	.	.	.	Fig-leaved goosefoot
Cirsium arvense	1	.	.	.	.	.	Creeping thistle
Echinochloa crus-galli	7	.	.	1	.	.	Barnyard grass
Fallopia convolvulus	296	8	2	23	.	.	Black bindweed
Fallopia dumetorum	.	15	.	.	.	.	Copse-bindweed
Galium aparine	2	1	1	.	.	.	Cleavers

Table 1 (continued). Absolute find numbers of charred archaeobotanical remains per site, section 2.

Section no./site name	2.1 Rosdorf-Mühlengrund	2.2 Rullstorf 8	2.3 Hittfeld 61	2.4 Lavenstedt 178	2.5 Sievern 114	2.6 Vechta 10	
Archaeological periodisation	LBK	EN II–MN II	FBC	MN I–II	Neol.	YN–LN	
Galium mollugo	2	.	.	.	.	.	False baby's breath
Galium spurium	13	1	.	.	.	.	False cleavers
Lapsana communis	8	.	.	.	.	.	Common nipplewort
Luzula luzuloides	1	.	.	.	.	.	Common woodrush
Persicaria lapathifolia	2	.	.	2	.	.	Curlytop knotweed
Persicaria maculosa	4	.	.	9	.	.	Lady's thumb
Phleum pratense/Poa annua	.	11	.	2	.	.	Timothy grass/Annual meadow grass
Plantago lanceolata	.	2	.	2	.	.	Ribwort plantain
Poa annua	1	.	.	.	.	.	Annual meadow grass
Polygonum aviculare	6	.	.	.	.	.	Knotweed
Rumex acetosella	1	.	.	.	.	.	Common sheep sorrel
Scleranthus annuus, inflorescences	.	2	.	1	.	.	Annual knawel
Setaria verticillata/viridis	1	.	.	.	.	.	Bristly foxtail/Green foxtail
Solanum nigrum	2	.	.	.	.	.	Black nightshade
Trifolium sp.	.	1	.	2	.	.	Clover
Sum wild taxa (dryland)	**514**	**45**	**3**	**51**	**0**	**0**	
Wild trees and shrubs (woodland)							
Alnus glutinosa	1	.	.	.	.	.	Common alder
Arctostaphylos uva-ursi	.	.	.	1	.	.	Common bearberry
Carpinus betulus	.	1	.	.	.	.	Common hornbeam
Fagus sylvatica, cupula	.	.	.	.	1	.	Common beech
Pinus sylvestris, cone fragments	.	118	.	.	.	.	Scotch pine
Tilia sp.	.	.	1	.	.	.	Lime tree
Sum wild trees and shrubs (woodland)	**1**	**119**	**1**	**1**	**1**	**0**	

Table 1 (continued). Absolute find numbers of charred archaeobotanical remains per site, section 2.

Section no./site name	2.1 Rosdorf-Mühlengrund	2.2 Rullstorf 8	2.3 Hittfeld 61	2.4 Lavenstedt 178	2.5 Sievern 114	2.6 Vechta 10	
Archaeological periodisation	LBK	EN II–MN II	FBC	MN I–II	Neol.	YN–LN	
Wetland/aquatic taxa							
Carex hirta	.	1	.	.	.	.	Hairy sedge
Sum wetland/aquatic taxa	**0**	**1**	**0**	**0**	**0**	**0**	
Varia							
Asteraceae p. p.	4	.	.	.	.	.	Composite family
Avena sp.	.	.	.	.	.	1	Oat
Brassicaceae p. p.	.	1	.	.	.	.	Cabbage family
Bromus sp.	93	1	.	.	.	.	Brome
Calluna sp., inflorescences	.	.	.	1	.	.	Heather
Carex sp., tricarpellat	.	1	.	.	.	.	Sedge, tricarpellate
Chenopodiaceae p. p.	2	.	.	.	.	.	Goosefoot family
Ericaceae, leaves	.	.	.	1	.	.	Heath family
Fabaceae p. p.	9	.	.	.	.	.	Legume family
Galium sp.	3	.	.	.	.	.	Bedstraw
Lathyrus/Pisum/Vicia	.	.	.	.	.	1	Peavine/Pea/Vetch
Lolium sp.	.	.	.	1	.	.	Ryegrass
Poaceae p. p.	8	4	.	7	.	.	Grass family
Polygonaceae p. p.	.	4	.	1	.	.	Knotweed family
Rumex sp.	.	1	.	.	.	.	Dock
Stellaria sp.	1	.	.	.	.	.	Stitchwort
Vicia sp.	.	281	.	4	1	.	Vetch
Sum varia	**120**	**293**	**0**	**15**	**1**	**2**	
Sum total (seeds/fruits)	**28685**	**477**	**219**	**990**	**7**	**220**	

Table 2. Absolute find numbers of charred archaeobotanical remains per site, section 3.

Section no./site name	3.1 Hemmingstedt LA2	3.2 Borgstedt LA35	3.3 Albersdorf LA5, Brutkamp	3.3 Albersdorf LA5, Brutkamp	3.4 Albersdorf LA68, Diecksknöll	
Archaeological periodisation	EN Ib–II	MN–LN	EN Ib	LN I	EN Ib–YN 2	
Crops						
Cereals, caryopses						
Triticum monococcum	.	1	.	.	.	Einkorn
Triticum dicoccon	5	13	.	.	1	Emmer
Triticum sp.	.	.	.	.	1	Wheat
Hordeum vulgare nudum	84	1	.	.	1	Free-threshing barley
Hordeum vulgare	137	5	1	1	5	Barley
Cerealia indet.	132	26	1	1	2	Cereals
Sum cereals, caryopses	**358**	**46**	**2**	**2**	**10**	
Cereals, threshing remains						
Triticum monococcum, glume bases	.	1	.	.	.	Einkorn
Triticum dicoccon, threshing remains	.	25	.	.	.	Emmer
Sum cereals, threshing remains	**0**	**26**	**0**	**0**	**0**	
Fruits/nuts/tubers						
Corylus avellana, nutshell fragments	10	8	48	23	53	Hazel
Ficaria verna, root tubers	.	.	.	.	11	Lesser celandine
Prunus spinosa	.	.	.	1	.	Blackthorn
Quercus sp., cupula	.	.	.	.	2	Oak
Rubus fruticosus	.	.	1	.	9	Bramble
Rubus idaeus	.	.	1	1	1	Raspberry
Rubus sp.	.	.	1	1	1	Bramble/Raspberry/Dewberry
Sum fruits/nuts/tubers	**10**	**8**	**51**	**26**	**77**	

Table 2 (continued). Absolute find numbers of charred archaeobotanical remains per site, section 3.

Section no./site name	3.1 Hemmingstedt LA2	3.2 Borgstedt LA35	3.3 Albersdorf LA5, Brutkamp	3.3 Albersdorf LA5, Brutkamp	3.4 Albersdorf LA68, Dieksknöll	
Archaeological periodisation	EN Ib–II	MN–LN	EN Ib	LN I	EN Ib–YN 2	
Wild taxa (dryland)						
Arrhenatherum elatius var. *bulbosum*, bulbs	.	1	.	15	.	Tuber oat grass
Calluna vulgaris, inflorescences	.	.	.	.	1	Heather
Chenopodium album	.	36	.	.	.	White goosefoot
Fallopia convolvulus	1	.	.	.	3	Black bindweed
Galium aparine	.	.	3	.	.	Cleavers
Galium spurium	.	1	.	.	.	False cleavers
Lapsana communis	.	.	1	.	.	Common nipplewort
Medicago sp.	.	.	.	.	1	Burclover
Persicaria lapathifolia	.	.	2	.	.	Curlytop knotweed
Persicaria lapathifolia/maculosa	.	14	.	.	.	Curlytop knotweed/Lady's thumb
Phleum pratense	.	.	.	.	1	Timothy grass
Plantago lanceolata	.	4	.	.	.	Ribwort plantain
Rumex acetosella	.	1	.	.	.	Common sheep sorrel
Sum wild taxa (dryland)	**1**	**57**	**6**	**15**	**6**	
Wild trees and shrubs (woodland)						
Carpinus betulus	.	.	.	.	1	Common hornbeam
Fagus sylvatica, cupula	.	.	.	.	4	Common beech
Tilia sp.	.	.	2	.	.	Lime
Sum wild trees and shrubs (woodland)	**0**	**0**	**2**	**0**	**5**	
Varia						
Allium sp., bulbs	.	.	.	.	1	Allium
Cyperaceae	.	.	.	1	.	Sedge family

Table 2 (continued). Absolute find numbers of charred archaeobotanical remains per site, section 3.

Section no./site name	3.1 Hemmingstedt LA2	3.2 Borgstedt LA35	3.3 Albersdorf LA5, Brutkamp	3.3 Albersdorf LA5, Brutkamp	3.4 Albersdorf LA68, Dieksknöll	
Archaeological periodisation	EN Ib–II	MN–LN	EN Ib	LN I	EN Ib–YN 2	
Equisetum sp., vegetative	.	.	.	.	8	Horsetail
Fabaceae p. p.	.	1	.	.	.	Legume family
Poaceae p. p.	.	2	.	.	.	Grass family
Polygonaceae p. p.	.	1	.	.	1	Knotweed family
Vicia sp.	1	1	.	.	4	Vetch
Sum varia	**1**	**5**	**0**	**1**	**14**	
Sum total (seeds/fruits)	**370**	**115**	**61**	**29**	**91**	

Table 3. Absolute find numbers of charred archaeobotanical remains per site, section 4.

Section no./site name	4.1 Rastorf LA6a/c	4.2 Rastorf LA73	4.3 Eisendorf LA42 and LA43	4.4 Stolpe-Depenau LA14	4.5 Bad Segeberg LA93	4.6 Bad Segeberg LA94	4.7 Bosau LA242	
Archaeological periodisation	EN II–MN I	EN Ib–EN II	EN–MN	FBC	EN II–MN II	EN Ib/II; MN I	LN	
Crops								
Cereals, caryopses								
Triticum monococcum	.	.	.	.	1	1	.	Einkorn
Triticum dicoccon	16	.	.	.	.	75	4	Emmer
Triticum aestivum s. l./*durum/turgidum*	12	.	.	.	.	1	5339	Free-threshing wheat
Triticum sp.	1	.	.	.	.	5	.	Wheat
Hordeum vulgare nudum	7	.	.	.	.	9	801	Free-threshing barley
Hordeum vulgare	28	.	2	.	.	76	1	Barley
Secale cereale	.	.	.	.	.	1	.	Rye
Cerealia indet.	167	1	1	1	1	378	.	Cereals
Sum cereals, caryopses	**231**	**1**	**3**	**1**	**2**	**546**	**6145**	
Cereals, threshing remains								
Triticum monococcum, glume bases	.	.	.	.	1	20	.	Einkorn
Triticum dicoccon, threshing remains	11	.	.	2	2	272	.	Emmer
Hordeum vulgare, rachis segments	.	.	.	.	.	1	.	Barley
Cerealia indet., threshing remains	.	.	.	.	.	1	.	Cereals
Sum cereals, threshing remains	**11**	**0**	**0**	**2**	**3**	**294**	**0**	
Oil plants								
Camelina sativa	.	.	.	.	.	1	.	Gold of pleasure
Linum usitatissimum	2	.	.	.	.	.	.	Linseed
Sum oil plants	**2**	**0**	**0**	**0**	**0**	**1**	**0**	
Fruits/nuts/tubers								
Cornus sanguinea	.	.	.	.	.	.	3	Bloodtwig dogwood

Table 3 (continued). Absolute find numbers of charred archaeobotanical remains per site, section 4.

Section no./site name	4.1 Rastorf LA6a/c	4.2 Rastorf LA73	4.3 Eisendorf LA42 and LA43	4.4 Stolpe-Depenau LA14	4.5 Bad Segeberg LA93	4.6 Bad Segeberg LA94	4.7 Bosau LA242	
Archaeological periodisation	EN II–MN I	EN Ib–EN II	EN–MN	FBC	EN II–MN II	EN Ib/II; MN I	LN	
Corylus avellana, nuts	10	.	.	.	.	.	.	Hazel
Corylus avellana, nutshell fragments	876	9	.	2	1	132	1	Hazel
Crataegus laevigata	.	.	.	.	.	1	.	Midland hawthorn
Ficaria verna, root tubers	18	.	.	.	.	48	.	Lesser celandine
Malus sylvestris, seeds	1	.	.	.	.	1	.	Crab apple
Malus sylvestris, flesh	94	2	.	.	.	17	.	Crab apple
Malus sylvestris, pedicels	.	.	.	.	.	1	.	Crab apple
Rosaceae p. p.	.	1	.	.	.	1	.	Rose family
Rubus fruticosus	8	.	13	.	.	3	.	Bramble
Rubus idaeus	.	.	.	.	.	1	.	Raspberry
Rubus sp.	1	.	.	.	.	.	.	Bramble/Raspberry/Dewberry
Rumex crispus	.	.	.	1	.	1	.	Curly dock
Sorbus aucuparia, flesh	.	.	.	.	.	1	.	Rowan
Sum fruits/nuts/tubers	**1008**	**12**	**13**	**3**	**1**	**207**	**4**	
Wild taxa (dryland)								
Aphanes arvensis	.	.	.	.	.	1	.	Parsley piert
Arrhenatherum elatius var. *bulbosum*, bulbs	.	.	.	.	.	2	.	Tuber oat grass
Bromus secalinus	.	.	.	.	.	1	.	Rye brome
Chenopodium album	.	1	.	.	.	36	.	White goosefoot
Chenopodium polyspermum	.	.	.	.	.	1	.	Many-seeded goosefoot
Chenopodium sp.	.	.	.	.	.	3	.	Goosefoot
Echinochloa crus-galli	.	.	.	.	.	1	.	Barnyard grass
Fallopia convolvulus	1	.	.	.	.	16	.	Black bindweed
Galeopsis segetum	.	.	.	.	.	1	.	Downy hemp nettle

Table 3 (continued). Absolute find numbers of charred archaeobotanical remains per site, section 4.

Section no./site name	4.1 Rastorf LA6a/c	4.2 Rastorf LA73	4.3 Eisendorf LA42 and LA43	4.4 Stolpe-Depenau LA14	4.5 Bad Segeberg LA93	4.6 Bad Segeberg LA94	4.7 Bosau LA242	
Archaeological periodisation	EN II–MN I	EN Ib–EN II	EN–MN	FBC	EN II–MN II	EN Ib/II; MN I	LN	
Galium aparine	13	1	1	.	.	8	.	Cleavers
Galium spurium	.	.	1	.	.	1	.	False cleavers
Persicaria lapathifolia	.	.	.	.		1	.	Curlytop knotweed
Persicaria lapathifolia/maculosa	.	.	.	.	1	.	.	Curlytop knotweed/Lady's thumb
Persicaria maculosa	.	.	.	.	.	7	.	Lady's thumb
Phleum pratense/Poa annua	.	.	.	.	.	100	.	Timothy grass/Annual meadow grass
Plantago lanceolata	1	.	.	.	.	1	.	Ribwort plantain
Rumex sanguineus	.	.	.	.	.	1	.	Redvein dock
Scleranthus annuus, inflorescences	.	.	.	.	.	1	.	Annual knawel
Stachys palustris	.	.	.	.	.	1	.	Marsh woundwort
Stellaria media	.	.	.	.	.	1	1	Common chickweed
Viola arvensis	.	.	.	.	.	1	.	Field pansy
Sum wild taxa (dryland)	**15**	**2**	**2**	**0**	**1**	**185**	**1**	
Wild trees and shrubs (woodland)								
Alnus glutinosa	.	.	.	.	.	.	.	Common alder
Alnus glutinosa, infructescences	2	.	.	.	.	.	.	Common alder
Tilia sp.	16	.	.	.	.	.	.	Lime tree
Sum wild trees and shrubs (woodland)	**18**	**0**	**0**	**0**	**0**	**0**	**0**	
Wetland/aquatic taxa								
Carex hirta	.	2	.	.	.	.	.	Hairy sedge
Sum wetland/aquatic taxa	**0**	**2**	**0**	**0**	**0**	**0**	**0**	
Varia								
Apiaceae p. p.	1	.	.	.	.	.	.	Parsley family

Table 3 (continued). Absolute find numbers of charred archaeobotanical remains per site, section 4.

Section no./site name	4.1 Rastorf LA6a/c	4.2 Rastorf LA73	4.3 Eisendorf LA42 and LA43	4.4 Stolpe-Depenau LA14	4.5 Bad Segeberg LA93	4.6 Bad Segeberg LA94	4.7 Bosau LA242	
Archaeological periodisation	EN II–MN I	EN Ib–EN II	EN–MN	FBC	EN II–MN II	EN Ib/II; MN I	LN	
Asteraceae p. p.	.	.	.	.	.	1	.	Composite family
Avena sp.	.	.	.	2	.	3	.	Oat
Brassicaceae p. p.	1	.	.	.	.	1	.	Cabbage family
Bromus sp.	.	.	.	.	.	1	.	Brome
Carex sp., tricarpellat	.	.	.	.	.	3	.	Sedge, tricarpellate
Chenopodiaceae p. p.	.	.	.	.	.	1	.	Goosefoot family
Cyperaceae p. p.	.	1	.	.	.	.	.	Sedge family
Galium sp.	.	.	1	.	.	.	.	Bedstraw
Lamiaceae p. p.	1	.	.	.	.	1	.	Labiate
Lamium sp.	1	.	.	.	.	.	.	Dead-nettle
Panicoideae p. p.	.	.	.	.	.	3	.	Panicoideae
Poaceae p. p.	.	1	.	.	.	184	.	Grass family
Polygonaceae p. p.	.	.	.	.	.	4	.	Knotweed family
Raphanus sp., resting buds	.	1	.	.	.	.	.	Radish
Silene sp.	1	.	.	.	.	.	.	Campion
Stachys sp.	.	.	1	.	.	.	.	Woundwort
Vicia sp.	9	2	.	1	.	10	.	Vetch
Sum varia	**14**	**5**	**2**	**3**	**0**	**212**	**0**	
Sum total (seeds/fruits)	**1174**	**19**	**20**	**7**	**4**	**1081**	**6150**	

Table 4. Absolute find numbers of charred archaeobotanical remains per site, section 5.

Section no./site name	5.1 Oldenburg LA77, Dannau	5.2 Oldenburg LA191	5.3 Oldenburg LA232	5.4 Wangels LA505	5.5 Grube LA65	5.6 Heringsdorf LA38, Süssau	
Archaeological periodisation	MN I–MN IV	EN II–MN IV	MN I–MN II	MN V	EN Ib–EN II	EN Ib; MN I/II	
Crops							
Cereals, caryopses							
Triticum monococcum	28	13	10	.	.	.	Einkorn
Triticum dicoccon	6778	3137	3166	19	2	2	Emmer
Triticum aestivum s. l./*durum/turgidum*	14	.	.	2	.	.	Free-threshing wheat
Triticum sp.	53	.	.	.	.	.	Wheat
Hordeum vulgare nudum	3393	5595	.	50	24	.	Free-threshing barley
Hordeum vulgare	3155	248	35	.	.	.	Barley
Secale cereale	.	.	3	.	.	.	Rye
Cerealia indet.	20846	316	10	27	50	11	Cereals
Sum cereals, caryopses	**34267**	**9309**	**3224**	**98**	**76**	**13**	
Cereals, threshing remains							
Triticum monococcum, glume bases	60	15	51	.	2	.	Einkorn
Triticum dicoccon, threshing remains	18229	781	911	419	1	.	Emmer
Triticum monococcum/dicoccon, glume bases	3	.	.	.	.	.	Einkorn/emmer
Triticum aestivum s. l./*durum/turgidum*, rachis segments	12	.	.	4	.	.	Free-threshing wheat
Triticum durum/turgidum, rachis segments	.	.	.	2	.	.	Durum wheat
Triticum sp., rachis segments	10	.	.	.	.	.	Wheat
Hordeum vulgare nudum, rachis segments	.	.	.	4	.	.	Free-threshing barley
Hordeum vulgare, rachis segments	1037	8	.	95	.	.	Barley
Cerealia indet., threshing remains	445	.	.	.	.	.	Cereals
Sum cereals, threshing remains	**19796**	**804**	**962**	**524**	**3**	**0**	
Oil plants							
Linum usitatissimum	1	.	.	.	.	.	Linseed

Table 4 (continued). Absolute find numbers of charred archaeobotanical remains per site, section 5.

Section no./site name	5.1 Oldenburg LA77, Dannau	5.2 Oldenburg LA191	5.3 Oldenburg LA232	5.4 Wangels LA505	5.5 Grube LA65	5.6 Heringsdorf LA38, Süssau	
Archaeological periodisation	MN I–MN IV	EN II–MN IV	MN I–MN II	MN V	EN Ib–EN II	EN Ib; MN I/II	
Papaver somniferum	1	.	.	2	.	.	Opium poppy
Sum oil plants	**2**	**0**	**0**	**2**	**0**	**0**	
Legumes							
Fabaceae (cultivated)	.	.	.	.	1	.	Pulses (cultivated)
Sum legumes	**0**	**0**	**0**	**0**	**1**	**0**	
Fruits/nuts/tubers							
Cornus sanguinea	1	.	.	.	.	.	Bloodtwig dogwood
Corylus avellana, nutshell fragments	1805	21	13	1	1	4	Hazel
Crataegus laevigata	2	.	.	.	.	.	Midland hawthorn
Crataegus monogyna	1	.	.	.	.	.	Single-seeded hawthorn
Crataegus sp.	2	.	.	.	.	.	Hawthorn
Malus sylvestris, seeds	611	6	.	.	.	.	Crab apple
Malus sylvestris, fruits	156	.	.	.	.	.	Crab apple
Malus sylvestris, flesh	537	.	.	1	2	.	Crab apple
Malus sylvestris, pedicels	7	.	.	.	.	.	Crab apple
Prunus sp., thorns	1	.	.	.	.	.	Stone fruit
Quercus sp., cupula	1	.	.	.	.	.	Oak
Rosa canina	1	.	.	.	.	.	Dog rose
Rosa sp.	16	.	.	.	.	.	Rose
Rubus fruticosus	27	.	.	1	.	.	Bramble
Rubus idaeus	19	2	.	.	.	.	Raspberry
Rubus sp.	6	.	.	.	.	.	Bramble/Raspberry/Dewberry
Rumex crispus	24	.	.	.	.	.	Curly dock
Rumex obtusifolius	3	.	.	.	.	.	Bitter dock

Table 4 (continued). Absolute find numbers of charred archaeobotanical remains per site, section 5.

Section no./site name	5.1 Oldenburg LA77, Dannau	5.2 Oldenburg LA191	5.3 Oldenburg LA232	5.4 Wangels LA505	5.5 Grube LA65	5.6 Heringsdorf LA38, Süssau	
Archaeological periodisation	MN I–MN IV	EN II–MN IV	MN I–MN II	MN V	EN Ib–EN II	EN Ib; MN I/II	
Sorbus aucuparia	1	.	.	.	.	.	Rowan
Sum fruits/nuts/tubers	**3221**	**29**	**13**	**3**	**3**	**4**	
Wild taxa (dryland)							
Anagallis sp.	2	.	.	.	.	.	Pimpernel
Aphanes arvensis	2	.	.	.	.	.	Parsley piert
Arrhenatherum elatius var. *bulbosum*, bulbs	1	.	.	.	.	.	Tuber oat grass
Bromus arvensis	98	.	.	.	.	.	Field brome
Bromus hordeaceus (ssp. *hordeaceus*)	6	.	.	.	.	.	Soft brome
Bromus secalinus	64	.	.	2	.	.	Rye brome
Bromus secalinus-type	.	3	.	.	.	.	Rye brome-type
Chenopodium album	1471	.	5	4	.	1	White goosefoot
Chenopodium ficifolium	1	.	.	.	.	.	Fig-leaved goosefoot
Chenopodium sp.	120	.	1	.	.	.	Goosefoot
Descurainia sophia	.	.	.	1	.	.	Flixweed
Echinochloa crus-galli	14	.	.	.	.	.	Barnyard grass
Fallopia convolvulus	33	2	.	.	.	1	Black bindweed
Festuca sp.	12	.	.	.	.	.	Fescue
Galium aparine	4	.	.	.	.	.	Cleavers
Galium spurium	1	.	.	.	.	.	False cleavers
Lapsana communis	1	.	.	.	.	.	Common nipplewort
Lolium temulentum	4	.	.	.	.	.	Darnel
Malva sp.	3	.	.	.	.	.	Mallow
Odontites vernus	2	.	.	.	.	.	Red Bartsia
Persicaria lapathifolia	3	.	.	.	.	.	Curlytop knotweed

Table 4 (continued). Absolute find numbers of charred archaeobotanical remains per site, section 5.

Section no./site name	5.1 Oldenburg LA77, Dannau	5.2 Oldenburg LA191	5.3 Oldenburg LA232	5.4 Wangels LA505	5.5 Grube LA65	5.6 Heringsdorf LA38, Süssau	
Archaeological periodisation	MN I–MN IV	EN II–MN IV	MN I–MN II	MN V	EN Ib–EN II	EN Ib; MN I/II	
Persicaria lapathifolia/maculosa	2	.	.	.	.	.	Curlytop knotweed/Lady's thumb
Persicaria maculosa	25	4	.	.	.	.	Lady's thumb
Phleum pratense	149	.	.	.	.	.	Timothy grass
Phleum pratense/Poa annua	77	.	.	.	.	.	Timothy grass/Annual meadow grass
Phleum sp.	.	.	6	13	.	.	Timothy
Plantago lanceolata	20	.	.	.	.	.	Ribwort plantain
Poa annua	40	.	.	.	.	.	Annual meadow grass
Polygonum aviculare	8	.	.	.	.	.	Knotweed
Rumex acetosella	5	.	.	.	.	.	Common sheep sorrel
Rumex sanguineus	1	1	.	.	.	.	Redvein dock
Setaria viridis	6	.	.	.	.	.	Green foxtail
Sherardia arvensis	1	.	.	.	.	.	Field madder
Solanum nigrum	46	.	.	.	.	.	Black nightshade
Stellaria media	5	1	.	.	.	.	Common chickweed
Trifolium sp.	42	.	.	.	.	.	Clover
Veronica hederifolia	.	.	.	.	.	.	Ivy-leaved speedwell
Sum wild taxa (dryland)	**2269**	**11**	**12**	**20**	**0**	**2**	
Wild trees and shrubs (woodland)							
Alnus glutinosa	2	.	.	1	.	.	Common alder
Alnus glutinosa, infructescences	1	.	.	.	.	.	Common alder
Sum wild trees and shrubs (woodland)	**3**	**0**	**0**	**1**	**0**	**0**	
Wetland/aquatic taxa							
Bolboschoenus sp.	.	.	.	1	.	.	Club rush
Carex disticha	1	.	.	.	.	.	Brown sedge

Table 4 (continued). Absolute find numbers of charred archaeobotanical remains per site, section 5.

Section no./site name	5.1 Oldenburg LA77, Dannau	5.2 Oldenburg LA191	5.3 Oldenburg LA232	5.4 Wangels LA505	5.5 Grube LA65	5.6 Heringsdorf LA38, Süssau	
Archaeological periodisation	MN I–MN IV	EN II–MN IV	MN I–MN II	MN V	EN Ib–EN II	EN Ib; MN I/II	
Carex hirta	12	.	.	.	.	.	Hairy sedge
Carex vulpina	2	.	.	.	.	.	True fox-sedge
Cladium mariscus	.	.	2	.	.	.	Great fen-sedge
Iris pseudacorus	.	.	.	1	.	.	Yellow iris
Juncus sp.	1	.	.	.	.	.	Rush
Mentha sp.	1	.	.	.	.	.	Mint
Persicaria hydropiper	3	.	.	.	.	.	Marsh-pepper smartweed
Persicaria minor	1	.	.	.	.	.	Pygmy smartweed
Potamogeton sp.	.	.	.	1	.	.	Pondweed
Schoenoplectus lacustris	.	1	.	.	.	.	True bulrush
Solanum dulcamara	.	5	.	.	.	.	Bittersweet
Trapa natans, pericarps	4	.	.	.	.	.	Water chestnut
Sum wetland/aquatic taxa	**25**	**6**	**2**	**3**	**0**	**0**	
Salt marshes							
Atriplex patula/prostrata	1	.	.	.	.	.	Common orache/Spear-leaved orache
Atriplex prostrata	3	.	.	.	.	.	Spear-leaved orache
Atriplex sp.	3	.	.	.	.	.	Orache
Ruppia maritima	.	.	.	1	.	.	Beaked tasselweed
Schoenoplectus sp.	3	.	.	.	.	.	Bulrush
Schoenoplectus tabernaemontani	3	.	.	1	.	.	Rough bulrush
Sum salt marshes	**13**	**0**	**0**	**2**	**0**	**0**	
Varia							
Alchemilla/Aphanes	1	.	.	.	.	.	Lady's mantle/Parsley piert
Allium sp.	2	.	.	.	.	.	Allium

Table 4 (continued). Absolute find numbers of charred archaeobotanical remains per site, section 5.

Section no./site name	5.1 Oldenburg LA77, Dannau	5.2 Oldenburg LA191	5.3 Oldenburg LA232	5.4 Wangels LA505	5.5 Grube LA65	5.6 Heringsdorf LA38, Süssau	
Archaeological periodisation	MN I–MN IV	EN II–MN IV	MN I–MN II	MN V	EN Ib–EN II	EN Ib; MN I/II	
Allium sp., bulbs	1	.	.	.	.	.	Allium
Anchusa sp.	1	.	.	.	.	.	Bugloss
Apiaceae p. p.	1	1	.	.	.	.	Parsley family
Asteraceae p. p.	3	.	.	.	.	.	Composite family
Brassicaceae p. p.	1	.	.	.	.	.	Cabbage family
Bromus sp.	17	.	1	.	.	.	Brome
Carex sp.	1	.	1	.	.	.	Sedge
Carex sp., bicarpellat	6	.	.	.	.	.	Sedge, bicarpellate
Carex sp., tricarpellat	5	.	.	.	.	.	Sedge, tricarpellate
Caryophyllaceae p. p.	6	.	.	.	.	.	Pink family
Chenopodiaceae p. p.	.	9	.	.	1	.	Goosefoot family
Cyperaceae p. p.	6	.	.	.	.	.	Sedge family
Danthonia sp.	1	.	.	.	.	.	Heath grass
Fabaceae p. p.	1	.	.	.	.	.	Legume family
Galium sp.	2	.	.	.	.	.	Bedstraw
Lamiaceae p. p.	5	.	.	.	.	.	Labiate
Lepidium sp.	1	.	.	.	.	.	Peppercress
Lolium sp.	6	.	.	.	.	.	Ryegrass
Panicoideae p. p.	6	.	.	.	.	.	Panicoideae
Poaceae p. p.	1280	58	1	5	24	.	Grass family
Poaceae, vegetative, p. p.	.	.	20	.	.	.	Grass family
Polygonaceae p. p.	10	.	.	.	.	.	Knotweed family
Ranunculaceae p. p.	2	.	.	.	.	.	Buttercup family
Ranunculus sp.	2	.	.	.	.	.	Buttercup
Solanaceae p. p.	1	.	.	.	.	.	Nightshade family

Table 4 (continued). Absolute find numbers of charred archaeobotanical remains per site, section 5.

Section no./site name	5.1 Oldenburg LA77, Dannau	5.2 Oldenburg LA191	5.3 Oldenburg LA232	5.4 Wangels LA505	5.5 Grube LA65	5.6 Heringsdorf LA38, Süssau	
Archaeological periodisation	MN I–MN IV	EN II–MN IV	MN I–MN II	MN V	EN Ib–EN II	EN Ib; MN I/II	
Vicia sp.	25	.	1	2	1	.	Vetch
Viola sp.	1	.	.	.	.	.	Violet
Sum varia	**1394**	**68**	**24**	**7**	**26**	**0**	
Sum total (seeds/fruits)	**40646**	**9423**	**3255**	**135**	**104**	**19**	

Table 5. Absolute find numbers of charred archaeobotanical remains per site, section 6.

Section no./site name	6.1 Friedrichsruhe, Fpl. 7, Fpl. 8	6.2 Pastitzer Forst, Fpl. 53	6.3 Groß Behnitz, Fdpl. 15	6.3 Groß Behnitz, Fdpl. 15	6.4 Wachow-Niebede, Fdpl. 4	6.5 Lietzow, Fdpl. 2	
Archaeological periodisation	F7: EN II–MN II F8: EN I; YN 1–2	EN Ib	FBC	LBK	LBK	LBK	
Crops							
Cereals, caryopses							
Triticum dicoccon	7	.	7	2	1	15	Emmer
Triticum monococcum/dicoccon	1	.	.	.	.	.	Einkorn/emmer
Triticum aestivum s. l./*durum/turgidum*	.	.	.	.	.	1	Free-threshing wheat
Triticum sp.	.	.	9	.	7	1	Wheat
Hordeum vulgare	5	2	.	.	1	.	Barley
Cerealia indet.	22	2	38	14	13	49	Cereals
Panicum miliaceum	1	1	.	.	1	.	Broomcorn millet
Sum cereals, caryopses	**36**	**5**	**54**	**16**	**23**	**66**	
Cereals, threshing remains							
Triticum monococcum, glume bases	.	.	40	1	.	2	Einkorn
Triticum dicoccon, threshing remains	2	.	1347	1	11	4	Emmer
Triticum monococcum/dicoccon, glume bases	2	.	150	.	10	8	Einkorn/emmer
Sum cereals, threshing remains	**4**	**0**	**1537**	**2**	**21**	**14**	
Oil plants							
Linum usitatissimum	.	.	.	.	11	.	Linseed
Sum oil plants	**0**	**0**	**0**	**0**	**11**	**0**	
Legumes							
Pisum sativum	1	.	.	.	2	2	Garden pea
Sum legumes	**1**	**0**	**0**	**0**	**2**	**2**	

Table 5 (continued). Absolute find numbers of charred archaeobotanical remains per site, section 6.

Section no./site name	6.1 Friedrichsruhe, Fpl. 7, Fpl. 8	6.2 Pastitzer Forst, Fpl. 53	6.3 Groß Behnitz, Fdpl. 15	6.3 Groß Behnitz, Fdpl. 15	6.4 Wachow-Niebede, Fdpl. 4	6.5 Lietzow, Fdpl. 2	
Archaeological periodisation	F7: EN II–MN II F8: EN I; YN 1–2	EN Ib	FBC	LBK	LBK	LBK	
Fruits/nuts/tubers							
Corylus avellana, nutshell fragments	5	13	.	.	.	1	Hazel
Crataegus sp.	1	.	.	.	.	.	Hawthorn
Quercus sp.	1	.	.	.	.	.	Oak
Rubus fruticosus	2	.	.	.	.	.	Bramble
Rubus idaeus	2	.	.	.	.	.	Raspberry
Rumex crispus	.	.	.	.	2	.	Curly dock
Sum fruits/nuts/tubers	**11**	**13**	**0**	**0**	**2**	**1**	
Wild taxa (dryland)							
Arrhenatherum elatius var. *bulbosum*, bulbs	1	.	.	.	.	.	Tuber oat grass
Bromus arvensis	.	1	5	.	.	.	Field brome
Bromus secalinus	.	.	.	.	7	.	Rye brome
Chenopodium album	.	.	.	.	15	1003	White goosefoot
Echinochloa crus-galli	.	.	.	.	1	.	Barnyard grass
Fallopia convolvulus	1	1	.	1	7	156	Black bindweed
Fallopia dumetorum	1	.	.	.	.	.	Copse-bindweed
Galium aparine	1	.	.	.	.	.	Cleavers
Persicaria lapathifolia	.	1	.	.	.	.	Curlytop knotweed
Persicaria maculosa	.	2	.	.	.	17	Lady's thumb
Polygonum aviculare	.	1	1	1	.	.	Knotweed
Rumex acetosella	2	.	.	.	.	2	Common sheep sorrel
Stipa sp., awn fragments	.	.	.	.	10	.	Feather grass

Table 5 (continued). Absolute find numbers of charred archaeobotanical remains per site, section 6.

Section no./site name	6.1 Friedrichsruhe, Fpl. 7, Fpl. 8	6.2 Pastitzer Forst, Fpl. 53	6.3 Groß Behnitz, Fdpl. 15	6.3 Groß Behnitz, Fdpl. 15	6.4 Wachow-Niebede, Fdpl. 4	6.5 Lietzow, Fdpl. 2	
Archaeological periodisation	F7: EN II–MN II F8: EN I; YN 1–2	EN Ib	FBC	LBK	LBK	LBK	
Thlaspi arvense	.	.	.	.	.	1	Field pennycress
Sum wild taxa (dryland)	**6**	**6**	**6**	**2**	**40**	**1179**	
Wild trees and shrubs (woodland)							
Carpinus betulus	2	.	.	.	.	.	Common hornbeam
Pinus sp., needle fragments	.	.	.	.	.	1	Pine
Sum wild trees and shrubs (woodland)	**2**	**0**	**0**	**0**	**0**	**1**	
Wetland/aquatic taxa							
Juncus sp.	.	.	.	.	.	1	Rush
Sum wetland/aquatic taxa	**0**	**0**	**0**	**0**	**0**	**1**	
Varia							
Caryophyllaceae p. p.	.	1	1	.	.	.	Pink family
Chenopodiaceae p. p.	.	1	.	.	.	.	Goosefoot family
Fabaceae p. p.	.	.	.	.	2	4	Legume family
Galium sp.	.	.	.	.	.	1	Bedstraw
Lamiaceae, vegetative	.	.	.	.	.	1	Labiate
Panicoideae p. p.	.	2	.	.	.	.	Panicoideae
Poaceae p. p.	1	.	.	.	2	.	Grass family
Poaceae, glume bases, p. p.	.	.	.	.	.	1	Grass family
Vicia sp.	1	.	.	.	.	.	Vetch
Sum varia	**2**	**4**	**1**	**0**	**4**	**7**	
Sum total (seeds/fruits)	57	28	61	18	72	1254	

Table 6. Absolute find numbers of charred archaeobotanical remains per site, section 7. (* incl. data by Jørgensen, 1977, 1982; ** incl. data by Høyem Andreasen, 2017. Full citation in the resp. section).

Section no./site name	7.1 Sarup FSM 50*	7.1 Sarup FSM 50*	7.1 Sarup FSM 50*	7.1 Sarup FSM 50*	7.2 Frydenlund OBM 6973**	7.3 Damsbo FSM 3527	7.3 Damsbo FSM 3527	7.3 Damsbo FSM 3527	7.3 Damsbo FSM 3527	7.4 Sneglebjerg FHM 9137	
Archaeological periodisation	EN II	EN II–MN IV	MN I–IV	MN V	EN Ib	EN II	MN I–IV	MN V	YN 1–2	MN Ib	
Crops											
Cereals, caryopses											
Triticum monococcum	3	9	33	.	28	.	.	.	2	.	Einkorn
Triticum dicoccon	946	3314	12783	.	276	2	.	14	2574	25	Emmer
Triticum monococcum/dicoccon	.	.	4	.	.	.	.	.	.	.	Einkorn/emmer
Triticum aestivum s. l./*durum/turgidum*	.	4	17	.	909	.	.	1	.	1	Free-threshing wheat
Triticum sp.	.	4	484	.	94	.	.	2	.	.	Wheat
Hordeum vulgare nudum	23	56	109	6	244	.	.	.	.	4	Free-threshing barley
Hordeum vulgare	.	34	133	.	210	4	.	5	.	6	Barley
Cerealia indet.	1672	9411	52407	16	4870	9	.	20	11600	164	Cereals
Cerealia indet., mesh	.	.	.	.	.	.	.	.	.	2	Cereals
Panicum miliaceum	.	.	.	.	1	.	.	.	.	.	Broomcorn millet
Sum cereals, caryopses	**2644**	**12832**	**65970**	**22**	**6632**	**15**	**0**	**42**	**14176**	**202**	
Cereals, threshing remains											
Triticum monococcum, glume bases	.	2	1	.	65	.	.	.	.	.	Einkorn
Triticum dicoccon, threshing remains	11	22	527	.	852	.	.	386	6711	1	Emmer
Triticum monococcum/dicoccon, glume bases	.	.	10	.	30	.	.	.	.	.	Einkorn/emmer
Triticum aestivum s. l./*durum/turgidum*, rachis segments	.	.	.	.	207	.	.	.	.	.	Free-threshing wheat
Triticum durum/turgidum, rachis segments	.	.	.	.	1263	.	.	.	.	.	Durum wheat
Triticum sp., "new-type", glume bases	.	.	.	.	3	.	.	.	.	.	Wheat, „new-type"
Triticum sp., glume bases	.	24	59	.	624	.	.	.	.	.	Glume wheat
Triticum sp., rachis segments	.	2	.	.	1623	.	.	.	.	.	Wheat
Hordeum vulgare nudum, rachis segments	.	.	.	.	3	.	.	.	.	.	Free-threshing barley

Table 6 (continued). Absolute find numbers of charred archaeobotanical remains per site, section 7.

Section no./site name	7.1 Sarup FSM 50*	7.1 Sarup FSM 50*	7.1 Sarup FSM 50*	7.1 Sarup FSM 50*	7.2 Frydenlund OBM 6973**	7.3 Damsbo FSM 3527	7.3 Damsbo FSM 3527	7.3 Damsbo FSM 3527	7.3 Damsbo FSM 3527	7.4 Sneglebjerg FHM 9137	
Archaeological periodisation	EN II	EN II–MN IV	MN I–IV	MN V	EN Ib	EN II	MN I–IV	MN V	YN 1–2	MN Ib	
Hordeum vulgare, rachis segments	.	.	8	.	10	.	.	.	.	.	Barley
Secale cereale, rachis segments	.	.	1	.	.	.	.	.	.	.	Rye
Cerealia indet., threshing remains	.	.	3	.	170	.	.	.	.	.	Cereals
Sum cereals, threshing remains	**11**	**50**	**609**	**0**	**4850**	**0**	**0**	**386**	**6711**	**1**	
Fruits/nuts/tubers											
Corylus avellana, nuts	.	.	.	.	1	4	.	.	.	.	Hazel
Corylus avellana, nutshell fragments	.	1873	1234	.	401	95	6	.	35	9	Hazel
Crataegus laevigata	.	.	.	.	.	4	.	.	.	.	Midland hawthorn
Crataegus sp.	.	.	.	.	.	4	.	.	.	.	Hawthorn
Ficaria verna, root tubers	.	.	.	.	394	21	1	.	2	.	Lesser celandine
Malus sylvestris, seeds	.	3	7	.	.	.	.	.	2	.	Crab apple
Malus sylvestris, flesh	.	4	128	.	.	3	.	.	35	.	Crab apple
Quercus sp., cupula	.	.	.	.	2	1	.	.	.	.	Oak
Rubus caesius	.	.	1	.	.	.	.	.	.	.	European dewberry
Rubus fruticosus	.	.	1	.	12	4	.	.	.	.	Bramble
Rubus idaeus	.	.	1	.	8	.	.	.	.	.	Raspberry
Rubus sp.	.	.	.	.	6	2	.	.	.	.	Bramble/Raspberry/Dewberry
Rumex obtusifolius	.	.	.	.	1	.	.	.	.	.	Bitter dock
Sum fruits/nuts/tubers	**0**	**1880**	**1372**	**0**	**825**	**138**	**7**	**0**	**74**	**9**	
Wild taxa (dryland)											
Arrhenatherum elatius var. *bulbosum*, bulbs	.	.	.	.	18	15	.	.	.	.	Tuber oat grass
Bromus secalinus	.	.	.	.	.	2	.	100	279	1	Rye brome
Bromus secalinus-type	.	.	.	.	.	.	.	.	.	.	Rye brome-type

Table 6 (continued). Absolute find numbers of charred archaeobotanical remains per site, section 7.

Section no./site name	7.1 Sarup FSM 50*	7.1 Sarup FSM 50*	7.1 Sarup FSM 50*	7.1 Sarup FSM 50*	7.2 Frydenlund OBM 6973**	7.3 Damsbo FSM 3527	7.3 Damsbo FSM 3527	7.3 Damsbo FSM 3527	7.3 Damsbo FSM 3527	7.4 Sneglebjerg FHM 9137	
Archaeological periodisation	EN II	EN II–MN IV	MN I–IV	MN V	EN Ib	EN II	MN I–IV	MN V	YN 1–2	MN Ib	
Chenopodium album	.	3	58	.	3	132	.	4	16	1	White goosefoot
Chenopodium sp.	.	.	14	.	3	.	.	.	.	.	Goosefoot
Fallopia convolvulus	.	.	30	.	11	4	.	.	8	.	Black bindweed
Festuca sp.	.	.	.	.	.	.	.	.	4	.	Fescue
Festuca/Lolium	.	.	.	.	1	.	.	.	.	.	Fescue/Ryegrass
Galeopsis bifida/tetrahit	.	.	.	.	.	.	.	.	2	.	Bifid hemp nettle/Common hemp nettle
Galeopsis sp.	.	.	.	.	2	.	.	.	.	.	Hemp nettle
Galium aparine	.	.	2	.	10	3	.	.	.	.	Cleavers
Galium spurium	.	.	.	.	.	2	.	.	.	.	False cleavers
Linum catharticum	.	.	.	.	1	.	.	.	.	.	Fairy flax
Medicago lupulina	.	.	.	.	1	1	.	.	.	.	Yellow clover
Medicago minima	.	.	.	.	4	.	.	.	.	.	Little bur-clover
Medicago sp.	.	.	.	.	.	1	.	.	.	.	Burclover
Melilotus/Medicago/Trifolium	.	.	.	.	9	.	.	.	.	.	Sweet clover/Burclover/Clover
Persicaria lapathifolia	.	.	6	.	.	1	.	.	.	.	Curlytop knotweed
Persicaria lapathifolia/maculosa	.	.	4	.	1	.	.	.	.	.	Curlytop knotweed/Lady's thumb
Persicaria maculosa	.	.	.	.	6	.	.	.	6	3	Lady's thumb
Phleum pratense	.	87	271	.	.	.	.	7	1400	.	Timothy grass
Plantago lanceolata	.	.	1	.	2	2	.	.	.	.	Ribwort plantain
Poa annua	.	.	265	.	.	.	.	.	.	.	Annual meadow grass
Poa cenisia	.	.	2	.	.	.	.	.	.	.	Mount Cenis meadow grass
Poa sp.	.	11	.	.	.	.	.	.	.	.	Meadow grass
Polygonum aviculare	.	3	2	.	3	1	.	.	.	.	Knotweed
Prunella vulgaris	.	.	2	.	.	.	.	.	.	.	Common self-heal

Table 6 (continued). Absolute find numbers of charred archaeobotanical remains per site, section 7.

Section no./site name	7.1 Sarup FSM 50*	7.1 Sarup FSM 50*	7.1 Sarup FSM 50*	7.1 Sarup FSM 50*	7.2 Frydenlund OBM 6973**	7.3 Damsbo FSM 3527	7.3 Damsbo FSM 3527	7.3 Damsbo FSM 3527	7.3 Damsbo FSM 3527	7.4 Sneglebjerg FHM 9137	
Archaeological periodisation	EN II	EN II–MN IV	MN I–IV	MN V	EN Ib	EN II	MN I–IV	MN V	YN 1–2	MN Ib	
Rumex acetosa	.	.	.	.	.	1	.	.	.	.	Common sorrel
Rumex acetosella	.	.	5	.	3	82	.	.	.	.	Common sheep sorrel
Rumex sanguineus	.	.	1	.	.	.	.	.	.	.	Redvein dock
Sanguisorba minor	.	1	.	.	.	.	.	.	.	.	Small burnet
Scleranthus annuus, inflorescences	.	.	1	.	.	.	.	.	.	.	Annual knawel
Silene dioica	.	.	1	.	.	.	.	.	.	.	Red campion
Silene latifolia subsp. *alba*	.	.	1	.	.	.	.	.	.	.	White campion
Solanum nigrum	.	.	1	.	.	.	.	.	.	1	Black nightshade
Spergula arvensis	.	.	1	.	.	.	.	.	.	.	Corn spurrey
Stachys annua	.	.	2	.	.	.	.	.	.	.	Annual woundwort
Stachys annua/arvensis	.	.	.	.	.	4	.	.	.	.	Annual woundwort/Field woundwort
Stachys arvensis	.	.	1	.	3	.	.	.	.	.	Field woundwort
Stellaria graminea/palustris	.	.	1	.	.	.	.	.	.	.	Grass-leaved stitchwort/Marsh stitchwort
Stellaria media	.	.	.	.	1	.	.	.	.	.	Common chickweed
Trifolium sp.	.	.	1	.	5	.	.	.	.	.	Clover
Veronica hederifolia	.	.	.	.	9	.	.	.	.	.	Ivy-leaved speedwell
Vicia sp., small-seeded	.	1	.	.	.	.	.	.	.	.	Vetch
Sum wild taxa (dryland)	**0**	**106**	**673**	**0**	**96**	**251**	**0**	**111**	**1715**	**6**	
Wild trees and shrubs (woodland)											
Alnus sp.	.	.	1	.	.	.	.	.	.	.	Alder
Salix sp., buds	.	.	1	.	.	.	.	.	.	.	Willow
Tilia sp.	.	.	.	.	95	34	4	.	.	.	Lime tree
Sum wild trees and shrubs (woodland)	**0**	**0**	**2**	**0**	**95**	**34**	**4**	**0**	**0**	**0**	

Table 6 (continued). Absolute find numbers of charred archaeobotanical remains per site, section 7.

Section no./site name	7.1 Sarup FSM 50*	7.1 Sarup FSM 50*	7.1 Sarup FSM 50*	7.1 Sarup FSM 50*	7.2 Frydenlund OBM 6973**	7.3 Damsbo FSM 3527	7.3 Damsbo FSM 3527	7.3 Damsbo FSM 3527	7.3 Damsbo FSM 3527	7.4 Sneglebjerg FHM 9137	
Archaeological periodisation	EN II	EN II–MN IV	MN I–IV	MN V	EN Ib	EN II	MN I–IV	MN V	YN 1–2	MN Ib	
Wetland/aquatic taxa											
Carex flacca/flava	.	.	1	.	.	.	.	.	.	.	Glaucous sedge/Yellow sedge
Carex muricata/vulpina	.	.	1	.	2	.	.	.	.	.	Rough sedge/True fox-sedge
Carex vulpina	.	.	.	.	1	.	.	.	.	.	True fox-sedge
Rumex conglomeratus	.	.	.	.	4	.	.	.	.	.	Clustered dock
Solanum dulcamara	.	.	.	.	.	.	.	.	4	.	Bittersweet
Sparganium erectum	.	.	1	.	.	.	.	.	.	.	Branched bur-reed
Sum wetland/aquatic taxa	**0**	**0**	**3**	**0**	**7**	**0**	**0**	**0**	**4**	**0**	
Salt marshes											
Atriplex sp.	.	.	.	.	1	.	.	.	.	.	Orache
Rumex maritimus	.	1	.	.	.	.	.	.	.	.	Golden dock
Sum salt marshes	**0**	**1**	**0**	**0**	**1**	**0**	**0**	**0**	**0**	**0**	
Varia											
Agrostis sp.	.	.	5	.	.	.	.	.	.	.	Bent
Arenaria serpyllifolia	.	1	1	.	.	.	.	.	.	.	Thyme-leaved sandwort
Avena sp.	.	.	.	.	4	.	.	.	.	.	Oat
Brassica sp.	.	.	.	.	1	.	.	.	.	.	Cabbage
Bromus sp.	.	.	2	.	9	.	.	20	80	1	Brome
Carex sp.	.	1	3	.	.	.	.	.	.	.	Sedge
Carex sp., bicarpellat	.	.	.	.	2	.	.	.	.	.	Sedge, bicarpellate
Carex sp., tricarpellat	.	.	.	.	1	2	.	.	.	.	Sedge, tricarpellate
Carlina vulgaris	.	1	.	.	.	.	.	.	.	.	Carline thistle
Caryophyllaceae p. p.	.	.	.	.	3	.	.	.	4	.	Pink family

Table 6 (continued). Absolute find numbers of charred archaeobotanical remains per site, section 7.

Section no./site name	7.1 Sarup FSM 50*	7.1 Sarup FSM 50*	7.1 Sarup FSM 50*	7.1 Sarup FSM 50*	7.2 Frydenlund OBM 6973**	7.3 Damsbo FSM 3527	7.3 Damsbo FSM 3527	7.3 Damsbo FSM 3527	7.3 Damsbo FSM 3527	7.4 Sneglebjerg FHM 9137	
Archaeological periodisation	EN II	EN II–MN IV	MN I–IV	MN V	EN Ib	EN II	MN I–IV	MN V	YN 1–2	MN Ib	
Cerastium sp.	.	.	.	.	1	.	.	.	.	.	Mouse ear
Cyperaceae p. p.	.	.	1	.	.	.	.	.	.	.	Sedge family
Fabaceae p. p.	.	.	.	.	100	4	.	.	.	.	Legume family
Galium sp.	.	.	1	.	24	4	.	.	.	.	Bedstraw
Hypericum sp.	.	.	.	.	2	.	.	.	.	.	St John's wort
Lamiaceae p. p.	.	.	1	.	15	.	.	.	2	.	Labiate
Lathyrus/Pisum/Vicia	.	.	.	.	7	.	.	.	.	.	Peavine/Pea/Vetch
Lolium sp.	.	.	.	.	.	.	.	.	8	.	Ryegrass
Panicoideae p. p.	.	.	.	.	1	.	.	.	.	.	Panicoideae
Persicaria sp.	.	.	1	.	1	.	.	.	.	.	Knotweed
Poaceae p. p.	.	1	296	.	296	1	.	13	800	1	Grass family
Polygonaceae p. p.	.	1	4	.	4	3	.	.	.	.	Knotweed family
Potentilla sp.	.	.	2	.	.	.	.	.	.	.	Cinquefoil
Ranunculus sp.	.	.	1	.	2	.	.	.	.	.	Buttercup
Rumex sp.	.	1	1	.	12	.	.	.	.	1	Dock
Stachys sp.	.	.	.	.	3	1	.	.	.	.	Woundwort
Veronica sp.	.	.	.	.	1	.	.	.	.	.	Speedwell
Vicia sp.	.	.	.	.	1	10	.	.	6	.	Vetch
Viola sp.	.	.	.	.	1	.	.	.	.	.	Violet
Sum varia	**0**	**6**	**319**	**0**	**491**	**25**	**0**	**33**	**900**	**3**	
Sum total (seeds/fruits)	**2644**	**14821**	**68209**	**22**	**7735**	**424**	**10**	**186**	**16832**	**220**	

Table 7. The waterlogged plant remains from the site Wangels LA505, Middle Neolithic V (Kroll, 2001; section 5.4).

Section no./site name	5.4 Wangels LA505, waterlogged	
Archaeological periodisation	MN V	
	Find numbers	
Crops		
Cereals, threshing remains		
Triticum dicoccon, glume bases	2731	Emmer
Hordeum vulgare, rachis segments	3	Barley
Sum cereals, threshing remains	**2734**	
Oil plants		
Papaver somniferum	21	Opium poppy
Sum oil plants	**21**	
Fruits/nuts		
Corylus avellana, nutshell fragments	30	Hazel
Crataegus laevigata	1	Midland hawthorn
Crataegus monogyna	2	Single-seeded hawthorn
Fragaria vesca	120	Wild strawberry
Malus sylvestris	95	Crab apple
Prunus sp.	1	Stone fruit
Rosa sp.	9	Rose
Rubus fruticosus	134	Bramble
Rubus idaeus	263	Raspberry
Sum fruits/nuts	**655**	
Tubers		
Daucus carota	2	Wild carrot
Sum tubers	**2**	
Wild taxa (dryland)		
Aethusa cynapium	1	Fool's parsley
Alliaria petiolata	1	Garlic mustard
Atriplex patula	448	Common orache
Atriplex sp.	267	Orache
Brassica nigra	42	Black mustard
Bromus arvensis/secalinus	6	Field brome/Rye brome
Chenopodium album	1633	White goosefoot
Chenopodium ficifolium	33	Fig-leaved goosefoot
Chenopodium glaucum/rubrum	1590	Glaucous goosefoot/Red goosefoot
Chenopodium sp.	8	Goosefoot
Cirsium sp.	22	Thistle
Descurainia sophia	35	Flixweed
Dianthus sp.	1	Pink
Epilobium hirsutum	5	Great hairy willow-herb
Eupatorium cannabinum	5	Hemp agrimony
Fallopia convolvulus	29	Black bindweed
Fallopia convolvulus/dumetorum	36	Black bindweed/Copse-bindweed
Galeopsis sp.	2	Hemp nettle
Hypericum perforatum	4	Common St John's wort
Juncus effusus	17	Soft rush
Lapsana communis	7	Common nipplewort
Leontodon autumnalis	2	Autumn hawkbit
Lychnis flos-cuculi	2	Ragged robin
Lythrum salicaria	78	Purple loosestrife
Persicaria lapathifolia	9	Curlytop knotweed

Table 7 (continued). The waterlogged plant remains from the site Wangels LA505, Middle Neolithic V (Kroll, 2001; section 5.4).

Section no./site name	5.4 Wangels LA505, waterlogged	
Archaeological periodisation	MN V	
	Find numbers	
Persicaria maculosa	53	Lady's thumb
Phleum sp.	4	Timothy
Plantago major	58	Broadleaf plantain
Poa sp.	18	Meadow grass
Polygonum aviculare	40	Knotweed
Ranunculus acris	11	Meadow buttercup
Ranunculus repens	14	Creeping buttercup
Rumex acetosella	10	Common sheep sorrel
Rumex crispus	17	Curly dock
Sanicula europaea	2	Wood sanicle
Saponaria officinalis	2	Soapwort
Solanum nigrum	4	Black nightshade
Sonchus arvensis	1	Field sow-thistle
Sonchus asper	5	Prickly sow-thistle
Stachys palustris	1	Marsh woundwort
Stellaria media	44	Common chickweed
Trifolium sp.	2	Clover
Urtica dioica	86	Stinging nettle
Valerianella dentata	7	Narrow-fruited cornsalad
Sum wild taxa (dryland)	**4662**	
Wild trees (woodland)		
Alnus glutinosa	135	Common alder
Betula sp.	9	Birch
Tilia sp.	1	Lime tree
Sum wild trees (woodland)	**145**	
Wetland/aquatic taxa		
Aster tripolium	7	Sea aster
Bolboschoenus sp.	36	Club rush
Carex disticha	1	Brown sedge
Carex hirta	46	Hairy sedge
Carex vulpina	7	True fox-sedge
Characeae, oogones	12411	Stonewort
Eleocharis ovata/uniglumis	2	Ovate spikerush/Creeping spikerush
Glaux maritima	23	Sea milkwort
Juncus bufonius	1587	Toad rush
Juncus gerardii	372	Saltmarsh rush
Lycopus europaeus	8	Gypsywort
Mentha sp.	638	Mint
Najas marina	126	Holly-leaved naiad
Phragmites australis	25	Common reed
Persicaria hydropiper	6	Marsh-pepper smartweed
Potamogeton sp.	205	Pondweed
Potentilla anserina	1	Silverweed
Rumex hydrolapathum	6	Great water dock
Salicornia europaea	4	Common glasswort
Samolus valerandi	9	Seaside brookweed
Solanum dulcamara	19	Bittersweet
Spergularia salina	1	Lesser sea-spurrey

Table 7 (continued). The waterlogged plant remains from the site Wangels LA505, Middle Neolithic V (Kroll, 2001; section 5.4).

Section no./site name	5.4 Wangels LA505, waterlogged	
Archaeological periodisation	MN V	
	Find numbers	
Stellaria palustris	7	Marsh stitchwort
Suaeda maritima	13	Annual sea-blite
Typha sp.	2	Reed mace
Zannichellia palustris	189	Horned pondweed
Sum wetland/aquatic taxa	**15751**	
Salt marshes		
Ruppia maritima	277	Beaked tasselweed
Schoenoplectus tabernaemontani	156	Rough bulrush
Sum salt marshes	**433**	
Varia		
Agrostis sp.	2	Bent
Apium sp.	1	Marshwort
Asteraceae p. p.	7	Composite family
Brassica sp.	1	Cabbage
Brassicaceae p. p.	5	Cabbage family
Brassicaceae, husk	1	Cabbage family
Carduus sp.	1	Thistle
Carex sp.	55	Sedge
Cerastium sp.	10	Mouse ear
Chenopodiaceae p. p.	1	Goosefoot family
Geraniaceae	1	Geranium family
Lamium sp.	2	Dead-nettle
Lolium sp.	2	Ryegrass
Montia sp.	21	Blink
Myosotis sp.	1	Forget-me-not
Poaceae p. p.	17	Grass family
Ranunculus sp.	78	Buttercup
Rorippa sp.	4	Yellow-cress
Rubiaceae p. p.	1	Bedstraw family
Rumex sp.	1	Dock
Silene sp.	20	Campion
Veronica sp.	1	Speedwell
Viola sp.	53	Violet
Sum varia	**286**	
Sum total (seeds/fruits, without oogones)	**9544**	

Table 8. New radiocarbon dates on charred plant material for the respective sites. Calibration was carried out with the CALIB Radiocarbon Calibration software (Stuiver et al., 2019).

Section no.	Site	Material	Feature no./ description	Sample no./ description	Sub-sample no.	Lab-Code	^{14}C-date BP	Standard deviation	Cal BCE, 2 sigma highest prob., Calib 7.1	Median probabi-lity cal BCE
2.2	Rullstorf 8	*Triticum dicoccon*	3322	3322	1	Poz-62904	4590	40	3510–3108	3360
2.2	Rullstorf 8	Cerealia indet.	5408	5408	1	Poz-65231	4480	40	3348–3082	3211
2.2	Rullstorf 8	*Carpinus* sp.	0/321/5	0/321/5:0	1	Poz-60647	4570	80	3522–3078	3278
2.2	Rullstorf 8	*Triticum dicoccon*, charred glume bases	5550	5550	1	Poz-62903	4495	35	3352–3089	3217
2.2	Rullstorf 8	*Corylus* sp., charred nutshell fragment	5282	5282	1	Poz-62905	4510	35	3256–3097	3214
2.2	Rullstorf 8	*Corylus* sp., charred nutshell fragment	5188	5188	1	Poz-65233	4600	35	3512–3329	3373
2.3	Hittfeld 61	*Panicum miliaceum*	84	27	1	Poz-97511	2410	30	548–401	483
2.4	Lavenstedt 178	*Corylus* sp., charred nutshell fragment	174	743	1	Beta-421914	4400	30	3098–2917	3013
2.4	Lavenstedt 178	*Corylus* sp., charred nutshell fragment	168	534	1	Beta-421913	4470	30	3338–3207	3222
2.6	Vechta 10	*Hordeum vulgare nudum*	126	126.1	1	Poz-62898	3495	35	1912–1739	1819
3.1	Hemmingstedt LA2	*Corylus* sp., charred nutshell fragment	208	208-1	1	KIA-50447	4915	30	3715–3644	3686
3.1	Hemmingstedt LA2	*Hordeum vulgare*	298	298-1	1	KIA-50448-1	4570	30	3374–3110	3338
3.1	Hemmingstedt LA2	*Hordeum vulgare*	298	298-1	2	KIA-50448-2	4590	30	3498–3436	3363
3.2	Borgstedt LA35	*Arrhenatherum elatius* var. *bulbosum*	8-OK	1	1	Poz-62908	3360	35	1701–1600	1654
3.3	Albersdorf LA5, Brutkamp	*Hordeum vulgare*	11005	10300	1	KIA-41839	2455	30	755–414	596
3.3	Albersdorf LA5, Brutkamp	*Corylus* sp., charred nutshell fragment	10303	10202	1	KIA-41837	4770	30	3640–3516	3571
3.3	Albersdorf LA5, Brutkamp	*Arrhenatherum elatius* var. *bulbosum*	11202	10162	1	KIA-42398	3715	35	2205–2019	2102
3.3	Albersdorf LA5, Brutkamp	*Arrhenatherum elatius* var. *bulbosum*	10102	10030	1	KIA-42399	3745	40	2234–2032	2154
3.3	Albersdorf LA5, Brutkamp	*Corylus* sp., charcoal	10402	10256	1	KIA-41838	4690	35	3529–3370	3452
3.3	Albersdorf LA5, Brutkamp	Cerealia indet.	11005	10308	1	KIA-41840	4745	35	3637–3379	3561
3.3	Albersdorf LA5, Brutkamp	Charcoal	10505	10310	1	KIA-41841	2480	30	773–482	635
3.4	Albersdorf LA68, Dieksknöll	*Fallopia convolvulus*	201 G	20363	1	KIA-46421	5190	55	4082–3932	4004
3.4	Albersdorf LA68, Dieksknöll	*Corylus* sp., charred nutshell fragment	202 A	20364	1	KIA-46424-3	4035	30	2627–2474	2541
3.4	Albersdorf LA68, Dieksknöll	*Rubus fruticosus*	202 D	20368	1	KIA-46427	4554	45	3373–3098	3225
3.4	Albersdorf LA68, Dieksknöll	cf. *Pinus* sp.	202 B1	20365	1	KIA-46425	8227	39	7357–7082	7245
3.4	Albersdorf LA68, Dieksknöll	cf. *Pinus* sp.	202 I	20665	1	KIA-46428	8245	40	7376–7138	7268
3.4	Albersdorf LA68, Dieksknöll	cf. *Pinus* sp.	201 G/H	20211	1	KIA-46422	8430	45	7581–7450	7512
3.4	Albersdorf LA68, Dieksknöll	cf. *Pinus* sp.	201 B	20210	1	KIA-46364	9595	55	9211–8798	8991
3.4	Albersdorf LA68, Dieksknöll	cf. *Pinus* sp.	201 I	20471	1	KIA-46423	8575	40	7657–7535	7590
3.4	Albersdorf LA68, Dieksknöll	*Corylus* sp., charred nutshell fragment	202 C	20367	1	KIA-46426	4371	32	3033–2907	2981
4.1	Rastorf LA6 a/c	*Corylus* sp., charred nutshell fragment	38	4	2	Poz-65227	4550	40	3371–3100	3216

Table 8 (continued). New radiocarbon dates on charred plant material for the respective sites. Calibration was carried out with the CALIB Radiocarbon Calibration software (Stuiver et al., 2019).

Section no.	Site	Material	Feature no./ description	Sample no./ description	Sub-sample no.	Lab-Code	^{14}C-date BP	Standard deviation	Cal BCE, 2 sigma highest prob., Calib 7.1	Median probabi-lity cal BCE
4.1	Rastorf LA6a/c	*Corylus* sp., charred nutshell fragment	38	1	1	Poz-65221	4590	30	3498–3331	3363
4.1	Rastorf LA6a/c	*Malus* sp., flesh	99	uppermost filling	2	Poz-65220	4590	40	3510–3108	3360
4.1	Rastorf LA6a/c	*Corylus* sp., charred nutshell fragment	99	uppermost filling	1	Poz-65219	4580	35	3498–3109	3350
4.1	Rastorf LA6a/c	*Malus sylvestris*, flesh, 1st half	99	lowermost filling	1	Poz-61049	4605	35	3514–3332	3432
4.1	Rastorf LA6a/c	*Triticum aestivum* s.l.	38	4	1	KIA-49765	5210	55	4180–3942	4025
4.1	Rastorf LA6a/c	*Corylus* sp., charred nutshell fragment	38	15	1	Poz-65229	4575	35	3497–3311	3340
4.1	Rastorf LA6a/c	*Triticum aestivum* s.l.	38	18	1	Poz-60535	4640	30	3515–3359	3461
4.1	Rastorf LA6a/c	*Corylus* sp., charred nutshell fragment	82	fire pit 1	1	Poz-65218	4540	40	3368–3099	3213
4.1	Rastorf LA6a/c	*Malus sylvestris*, flesh, 2nd half	99	lowermost filling	2	Poz-62986	4710	40	3632–3372	3495
4.1	Rastorf LA6a/c	*Corylus* sp., charred nutshell fragment	38	16	1	Poz-65230	4575	35	3497–3105	3340
4.1	Rastorf LA6a/c	*Triticum* sp.	38	18	2	Poz-62985	4660	40	3524–3361	3454
4.2	Rastorf LA73	*Corylus* sp., charred nutshell fragment	trench 4, eastern part	ditch end, center	1	Poz-63038	4740	35	3636–3498	3554
4.2	Rastorf LA73	*Corylus* sp., charred nutshell fragment	trench 4, eastern part	uppermost filling	1	Poz-63039	4665	35	3522–3364	3452
4.2	Rastorf LA73	*Corylus* sp., charred nutshell fragment	trench 4, eastern part	uppermost filling	1	Poz-63037	4670	30	3521–3367	3450
4.2	Rastorf LA73	*Corylus* sp., charred nutshell fragment	pit 4	lowermost humic layer	1	Poz-63040	4670	40	3526–3363	3453
4.6	Bad Segeberg LA94	*Corylus* sp., charcoal	1-Fl3-Q3	BP 121-Pl 19	1	Poz-61026	4690	40	3532–3368	2457
4.6	Bad Segeberg LA94	*Hordeum vulgare*	1-Fl3-Q3	BP 84	1	Poz-65235	4485	35	3346–3088	3218
4.6	Bad Segeberg LA94	*Corylus* sp., charred nutshell, 1st half	1-Fl2-Q3	BP 11	1	Poz-61023	4535	35	3363–3102	3210
4.6	Bad Segeberg LA94	*Corylus* sp., charred nutshell, 2nd half	1-Fl2-Q3	BP 11	2	Poz-62913	4600	40	3516–3325	3373
4.6	Bad Segeberg LA94	*Hordeum vulgare*	1-Fl3-Q3	BP 60	1	Poz-62912	4500	35	3354–3090	3216
4.6	Bad Segeberg LA94	*Corylus* sp., charred nutshell fragment	1-Fl2-Q3	BP 9	1	Poz-65234	4655	35	3520–3361	3454
4.6	Bad Segeberg LA94	*Ficaria verna*, tuber	1-Fl3-Q3	BP 109-Pl 14	1	Poz-62914	4720	40	3634–3374	3512
4.6	Bad Segeberg LA94	*Corylus* sp., charred nutshell fragment	1-Fl3-Q3	BP 101-Pl 11	1	Poz-61025	4515	35	3356–3098	3212
4.7	Bosau LA242	*Triticum aestivum* s.l.	Pit	Bosau-Grube	1	Poz-62910	3595	35	2036–1879	1952

Table 8 (continued). New radiocarbon dates on charred plant material for the respective sites. Calibration was carried out with the CALIB Radiocarbon Calibration software (Stuiver et al., 2019).

Section no.	Site	Material	Feature no./ description	Sample no./ description	Sub-sample no.	Lab-Code	^{14}C-date BP	Standard deviation	Cal BCE, 2 sigma highest prob., Calib 7.1	Median probability cal BCE
5.1	Oldenburg LA77, Dannau	*Triticum dicoccon*	4034	4106	1	Poz-65324	4340	35	3026–2894	2961
5.1	Oldenburg LA77, Dannau	*Hordeum vulgare*	S1c_29	252	1	Poz-62784	4450	40	3338–3008	3152
5.1	Oldenburg LA77, Dannau	*Corylus* sp., charred nutshell fragment	5047	5175	1	Poz-65327	4425	35	3120–2922	3063
5.1	Oldenburg LA77, Dannau	*Hordeum vulgare*	S1c_23	258	1	Poz-65323	4305	30	2943–2883	2912
5.1	Oldenburg LA77, Dannau	*Triticum dicoccon*	3035	3109	1	Poz-65322	4280	30	2930–2873	2900
5.1	Oldenburg LA77, Dannau	*Triticum dicoccon*	7013	7024	1	Poz-65325	4385	35	3095–2910	2994
5.1	Oldenburg LA77, Dannau	*Triticum dicoccon*	6036	6002	1	Poz-65326	4380	30	3035–2913	2985
5.1	Oldenburg LA77, Dannau	*Hordeum vulgare*	1012	1048	1	Poz-62788	4440	35	3332–3002	3097
5.1	Oldenburg LA77, Dannau	*Corylus* sp., charred nutshell fragment	4037	4096	1	Poz-62792	4390	40	3105–2904	3005
5.1	Oldenburg LA77, Dannau	*Triticum dicoccon*	4056	4122	1	Poz-62794	4410	40	3118–2913	3039
5.1	Oldenburg LA77, Dannau	*Hordeum vulgare*	4008-c	4114	1	Poz-62793	4335	35	3025–2892	2956
5.1	Oldenburg LA77, Dannau	*Triticum dicoccon*	5007-5045	5155	1	Poz-62791	4380	40	3099–2903	2993
5.1	Oldenburg LA77, Dannau	*Triticum dicoccon*	6020	6001	1	Poz-62786	4430	40	3330–2922	3078
5.1	Oldenburg LA77, Dannau	*Triticum dicoccon*	1012	1057	1	Poz-62790	4445	35	3334–3007	3114
5.1	Oldenburg LA77, Dannau	*Hordeum vulgare*	6092	6042	1	Poz-65015	4405	35	3109–2912	3025
5.1	Oldenburg LA77, Dannau	*Hordeum vulgare*	1012	1028	1	Poz-62787	4450	35	3337–3009	3154
5.1	Oldenburg LA77, Dannau	*Malus* sp.	10434	10257	1	Poz-62785	4480	40	3348–3082	3211
5.1	Oldenburg LA77, Dannau	*Triticum dicoccon*	5007_sunken-floor	5154	1	Poz-62783	4560	60	3381–3088	3245
5.1	Oldenburg LA77, Dannau	*Malus* sp.	1012	1056	1	Poz-65021	4385	35	3095–2910	2994
5.1	Oldenburg LA77, Dannau	*Triticum dicoccon*	1012	1054	1	Poz-65022	4420	35	3118–2920	3054
5.1	Oldenburg LA77, Dannau	*Triticum dicoccon*	10503	10297	1	Poz-65017	4380	35	3091–2911	2989
5.1	Oldenburg LA77, Dannau	*Hordeum vulgare*	S1a_35	324	1	Poz-65016	4375	35	3039–2909	2985
5.2	Oldenburg LA191	*Hordeum vulgare*	3103	86/583	1	Poz-65217	4395	35	3102–2910	3008
5.2	Oldenburg LA191	*Hordeum vulgare*	3003	86/195	1	Poz-62899	4465	35	3339–3022	3198
5.3	Oldenburg LA232	*Triticum dicoccon*	3012	3011	3011	Poz-88422	4505	35	3352–3095	3215
5.3	Oldenburg LA 232	*Triticum dicoccon*	3020	3020	3020	Poz-88425	4455	35	3339–3011	3171
5.3	Oldenburg LA232	*Triticum dicoccon*	3016	3015	3015	Poz-88423	4445	35	3334–3007	3114
5.3	Oldenburg LA232	*Triticum dicoccon*	3011	3010	3010	Poz-88421	4450	35	3337–3009	3154
5.4	Wangels LA505	*Triticum dicoccon*	381	sherd inclusion	1	Poz-62902	4955	35	3797–3652	3713
5.4	Wangels LA505	*Hordeum vulgare*	domestic	4	1	Poz-65238	4240	40	2919–2737	2869
5.4	Wangels LA505	*Triticum dicoccon*, partly charred glume bases	domestic	2	1	Poz-65239	4305	35	3012–2882	2915

Table 8 (continued). New radiocarbon dates on charred plant material for the respective sites. Calibration was carried out with the CALIB Radiocarbon Calibration software (Stuiver et al., 2019).

Section no.	Site	Material	Feature no./ description	Sample no./ description	Sub-sample no.	Lab-Code	^{14}C-date BP	Standard deviation	Cal BCE, 2 sigma highest prob., Calib 7.1	Median probability cal BCE
5.5	Grube LA65	*Hordeum vulgare nudum*	3	06	1	Poz-62907	4790	40	3652–3515	3570
5.6	Heringsdorf LA38, Süssau	Cerealia indet.	St. 56	56	1	Poz-65236	4490	40	3350–3086	3214
5.6	Heringsdorf LA38, Süssau	*Corylus* sp., charred nutshell fragment	St. 40	40	1	Poz-65237	4800	35	3651–3521	3569
6.1	Friedrichsruhe, Fpl. 8	Cerealia indet.	S2_14	1	1	KIA-49747	5095	30	3965–3800	3864
6.1	Friedrichsruhe, Fpl. 8	*Carpinus* sp.	S3_5	8	1	Poz-60644	1240	30	AD 684–877	AD 759
6.1	Friedrichsruhe, Fpl. 8	Cerealia indet.	S2_8	2	1	KIA-49746	4804	25	3647–3526	3563
6.1	Friedrichsruhe, Fpl. 8	cf. *Pisum* sp.	S2_16	13	1	KIA-49745	4800	30	3648–3523	3567
6.1	Friedrichsruhe, Fpl. 8	*Arrhenatherum elatius* var. *bulbosum*	S3_5	6	1	KIA-49748	4047	30	2635–2476	2568
6.2	Pastitzer Forst, Fpl. 53	*Corylus* sp., charred nutshell fragment	10, palaeosol below long barrow	1	1	Poz-92792	4970	40	3804–3654	3746
6.2	Pastitzer Forst, Fpl. 53	*Corylus* sp., charred nutshell fragment	25, burial mound	4	1	Poz-92793	4860	40	3711–3628	3654
6.2	Pastitzer Forst, Fpl. 53	*Hordeum vulgare*	22, inner stone row	7	1	Poz-92812	2770	35	1001–835	915
7.1	Sarup FSM 50	*Triticum dicoccon*	A1097	39772	1	KIA-49236	4646	25	3514–3423	3461
7.1	Sarup FSM 50	*Triticum dicoccon*	A2005	39768	2	KIA-49235	4790	30	3610–3521	3461
7.1	Sarup FSM 50	*Triticum dicoccon*	A851	39777	1	KIA-49237	4619	25	3499–3352	3468
7.1	Sarup FSM 50	*Triticum dicoccon*	A1744	39767	1	KIA-49234	4890	30	3711–3638	3674
7.1	Sarup FSM 50	*Triticum dicoccon*	A1655	39763	1	KIA-49233	4725	30	3633–3377	3522
7.1	Sarup FSM 50	*Triticum dicoccon*	A2873	31447	1	KIA-49232	4855	35	3705–3631	3650
7.2	Frydenlund OBM 6973	*Triticum durum/turgidum*, rachis segments	A52	FS-91_6334	1	KIA-49229	5063	29	3953–3793	3871
7.2	Frydenlund OBM 6973	*Ficaria verna*, tuber	A18	FS-154_6227	1	Poz-62917	5180	40	4055–3937	3990
7.2	Frydenlund OBM 6973	*Triticum* sp.	A18	FS-153_6269	1	Poz-62983	4820	35	3606–3522	3576
7.2	Frydenlund OBM 6973	*Triticum durum/turgidum*, rachis segments	A52	FS-91_6227	1	KIA-49228	5035	35	3949–3760	3863
7.2	Frydenlund OBM 6973	*Hordeum vulgare nudum*	A18	FS-154_6256	1	Poz-62915	4840	40	3703–3526	3639
7.3	Damsbo FSM 3527	*Ficaria verna*, tuber	121	8814	1	Poz-62911	4640	40	3521–3350	3455

Citation: Stuiver, M., Reimer, P. J., and Reimer, R. W., 2019. CALIB 7.1 [WWW program] at http://calib.org, accessed 2019-4-17.

Index 1: Scientific plant names

Index 2: English plant names

Index 3: German plant names